AF874755

Principles of Radiopharmacology

DEVELOPMENTS IN NUCLEAR MEDICINE
Series editor Peter H. Cox

Cox, P. H. (ed.): Cholescintigraphy. 1981. ISBN 90-247-2524-0
Cox, P. H. (ed.): Progress in radiopharmacology 3. Selected Topics. 1982. ISBN 90-247-2768-5
Jonckheer, M. H., and Deconinck, F. (eds.): X-ray fluorescent scanning of the thyroid. 1983. ISBN 0-89838-561-X
Kristensen, K. and Nørbygaard, E. (eds.): Safety and efficacy of radiopharmaceuticals. 1984. ISBN 0-89838-609-8
Bossuyt, A. and Deconinck, F.: Amplitude/phase patterns in dynamic scintigraphic imaging. 1984. ISBN 0-89838-641-1
Hardeman, M. R. and Najean, Y. (eds.): Blood cells in nuclear medicine I. Cell kinetics and bio-distribution. 1984. ISBN 0-89838-653-5
Fueger, G. F. (ed.): Blood cells in nuclear medicine II. Migratory blood cells. 1984. ISBN 0-89838-654-3
Biersack, H. J. and Cox, P. H. (eds.): Radioisotope studies in cardiology. 1985. ISBN 0-89838-733-7
Cox, P. H., Limouris, G. and Woldring, M. G. (eds.): Progress in radiopharmacology 1985. 1985. ISBN 0-89838-745-0
Cox, P. H., Mather, S. J., Sampson, C. B. and Lazarus, C. R. (eds.): Progress in radiopharmacy 1986. 1986. ISBN 0-89838-823-6
Deckart, H. and Cox, P. H. (eds.): Principles of radiopharmacology. 1987. ISBN 0-89838-774-4

Principles of Radiopharmacology

edited by

Harald Deckart

Nuclear Medicine Clinic, Berlin-Buch, German Democratic Republic

Peter H. Cox

Department of Nuclear Medicine, Rotterdamsch
Radio-Therapeutisch Instituut, Rotterdam, The Netherlands

1987 SPRINGER-SCIENCE+BUSINESS MEDIA, B.V.

Book information

ISBN 978-94-010-8384-3   ISBN 978-94-009-4245-5 (eBook)
DOI 10.1007/978-94-009-4245-5

Library of Congress Cataloging-in-Publication Data
Principles of Radiopharmacology.
Based on training courses organized by the European Joint Committee on Radiopharmaceuticals.
1. Radiopharmaceuticals. I. Deckart, Harald. II. Cox, Peter H. III. European Joint Committee on Radiopharmaceuticals. [DNLM: 1. Nuclear Medicine. 2. Radioisotopes. WN 440 P9565]
RM852.P74 1986 615.8'424 85-26008

Copyright

© 1987 by Springer Science+Business Media Dordrecht
Originally published by Martinus Nijhoff Publishers, Dordrecht in 1987
Softcover reprint of the hardcover 1st edition 1987
All rights reserved. No part of this publication may be reproduced, stored in a retrieval system, or transmitted in any form or by any means, mechanical, photocopying, recording, or otherwise, without the prior written permission of the publishers,
Springer-Science+Business Media, B.V.

Contributors

Beyer, Gerd-Juergen, PhD, Professor of Chemistry
Central Institute for Reactor Research, Department of Radioactive Isotopes, GDR Academy of Sciences, Dresden-Rossendorf (GDR)

Blottner, Annemarie, Dr., Dipl. biol.
Nuclear Medicine Clinic, Department of Experimental Research, Klinikum Berlin-Buch (GDR)

Cox, Peter H., BSc PhD, MPS, ACT (Birm), Dipl. pharm.
Coordinating Chairman of the European Joint Committee on Radiopharmaceuticals. Rotterdamsch Radiotherapeutisch Instituut Dr. Daniel den Hoed Kliniek, Rotterdam (The Netherlands)

Deckart, Harald, MD, PhD, Professor of Nuclear Medicine
Academy of Postgraduate Medical Education and Nuclear Medicine Clinic, Klinikum Berlin-Buch (GDR)

Frier, Malcolm, BSc PhD, MPS Dipl. pharm.
Queen's Medical Centre, Department of Medical Physics, Radiopharmacy Unit. Nottingham–England (UK)

Jánoki, Győző A., PhD, Dipl. pharm.
'Frédéric Joliot-Curie' National Research Institute of Radiobiology and Radiohygiene. Department of Applied Isotopes, Budapest (Hungary)

Johannsen, Bernd, PhD, Reader, Dipl. chem.
Nuclear Medicine Clinic, University Hospital (Charité), Berlin (GDR)

Kaempfer, Ingrid, Dr., Dipl. chem.
Radiological Clinic, Department of Nuclear Medicine, University Hospital, Leipzig (GDR)

Kristensen, Knud, PhD, Dipl. pharm.
The Isotope Pharmacy–WHO Collaborating Centre for Nuclear Medicine (Radiopharmaceuticals), Copenhagen (Denmark)

Krumbiegel, Peter, Dr., Dipl. chem.
Central Institute for Isotope and Radiation Research, Department of Tracer Research, GDR Academy of Sciences, Leipzig (GDR)

Muenze, Rudolf, PhD, Professor of Chemistry
Central Institute for Reactor Research, Department of Radioactive Isotopes, GDR Academy of Sciences, Dresden-Rossendorf (GDR)

Pfeiffer, Gertrude, Dr., Dipl. chem.
Mallinckrodt Diagnostica, Wien (Austria)

Schmidt, Eberhardt H., Dr., Dipl. chem.
Central Institute for Isotope and Radiation Research, Department of Radiation Sources and Radiopharmaceuticals, GDR Academy of Sciences, Berlin-Buch (GDR)

Tautz, Manfred, PhD, Dipl. phys.
Institute for Clinical Radiation Physics, Klinikum Berlin-Buch (GDR)

Vormum, Guenter K., PhD, Professor of Chemistry
Central Institute for Isotope and Radiation Research, Department of Radiation Sources and Radiopharmaceuticals, GDR Academy of Sciences, Berlin-Buch (GDR)

Contents

Preface

This publication contains a collection of 22 manuscripts by authors invited to write review articles. They are based on lectures presented on the First and Second Training Course in Radiopharmacy and Radiopharmacology. These courses were organized under the auspices of the "Joint Committee on Radiopharmaceuticals" of both European Societies of Nuclear Medicine (ENMS and SNME) and were sponsored by the European Society of Nuclear Medicine (ENMS).
Recent developments in radiopharmacy and radiopharmacology have been very complex; they cannot be overlooked by the medical staff in nuclear medical departments. This book has been written to provide access to recent research and to practical daily routine in patients' investigations.

It provides a wide-ranging and topical overview of radiopharmacology including chapters on basic chemistry and physics, incorporation dosimetry, interference of drugs in radiopharmacokinetics, legal aspects and stable isotope-labelled pharmaceuticals. Style of presentation is not uniform. Partial overlapping concerning the content of papers has not been avoided in order to ensure different views and aspects of the same subject. We hope that the interdisciplinary approach will be stimulating and thought-provoking for the reader. Consequently, this book is designed for specialists working in nuclear medical centres which involve different disciplines such as pharmacists, radiochemists, physicists, biochemists, biologists, mathematicians, electronic engineers, physicians with different specialities, and technicians.

Harald Deckart – Peter H. Cox

Acknowledgments

The editors would like to thank *Ernst Strehlau,* Dipl. chem., for providing editorial assistance, which was of great help in compiling all articles and the bibliography; *Peter J. Luecke,* Dipl. phil., who undertook the important, difficult, but generally underrated job of checking all English manuscripts and translating several articles; Mrs. *Erika Wilczynski,* who conscientiously typed all manuscripts.

We particularly thank *Bernd Rolle,* Gustav Fischer Publishers, Jena, for his excellent assistance and commitment.

Finally, thanks are due to the contributors for their keen interest in writing this book. May all these efforts be of good use for stimulating nuclear medical practice.

Introduction

With the introduction of radiotracers into the life sciences by George von Hevesy in 1923 a new approach was made to label drugs with radioactive isotopes or with stable isotopes (Rudolf Schoenheimer 1939), called nuclear pharmaceuticals or mostly radiopharmaceuticals. In general, they do not have any pharmacodynamic effects on the organism. During the last 3 decades there has been a tremendous development of new labelled compounds for diagnostic applications (γ-emitters) and for treatment (β-emitters).

Radiochemistry is the science concerned with preparation techniques of radiotracers, activating isotopes by reactors or accelerators, their separation from impurities and research of structure-binding-relationship (Chapters 1, 2, 7).

Radiopharmacy is the production of radiopharmaceuticals, in particular the technique of labelling the radioactive isotope to the chemical compound (drug); investigation of in vitro- and in vivo-stability, and quality control (Chapters 3, 4, 6, 7, 9–11, 14–17).

Radiopharmacology studies the biological distribution of radioactive drugs (i. e. Radiopharmacokinetics, Chapters 12, 13, 20) the interactions of radiotracers or -pharmaceuticals with binding sites, mechanism of transport, binding to cells, the fate of metabolites and the medical applications (Chapters 8, 12–14, 18–20, 22).

Chapter 1

Fundamentals of Radioactivity

Guenter K. Vormum

1. The atomic nucleus, nuclear systematics, rules for radioactive transformation

Atoms consist of the negatively-charged electron sheath and the nucleus which carries positive charge. The number of electrons in the sheath equals the positive charges in the nucleus. The diameter of an atom is about 10^{-8} cm while that of the nucleus is only 10^{-13} to 10^{-12} cm.

The *atomic nucleus* consists of protons and neutrons (with the exception of the nucleus of the common hydrogen atom). These particles are collectively called nucleons. Protons have a positive charge equal in magnitude to the negative charge on an electron ($4.8030 \cdot 10^{-10}$ electrostatic units). Neutrons, on the other hand, have no charge.

The mass of the nucleons is approximately equal to that of a hydrogen atom and therefore nearly equal to unity on the atomic weight scale which is based on the mass of an atom of ^{12}C taken as exactly 12.0000 units.

The masses of atomic nuclei as well as (indirectly) that of the neutron have been determined by means of mass spectrometric methods with high precision (0.01 to 10 ppm).

On the ^{12}C scale the *mass of a*

hydrogen atom $M_H = 1.0078252$ and that of a
neutron $M_n = 1.0086654$ mass units.
One mass unit equals $1.660 \cdot 10^{-24}$ g.

One consequence of *Einstein's* special Theory of Relativity is the equivalence of mass and energy. The total energy content E of a system of mass M is given by the relation

$$E = Mc^2 \tag{1}$$

(c is the velocity of light $= 2.99792 \cdot 10^{10}$ cm/s)

Thus it can be calculated that
one a.m.u. $= 1.492 \cdot 10^{-3}$ erg

In nuclear science, energy is commonly expressed in units of electron volts (eV), kiloelectron volts (keV) or million electron volts (MeV).

1 eV is the energy required to raise one electron through a potential difference of 1 volt.

1 MeV	$= 1.602.10^{-6}$ erg and hence
1 a.m.u.	= 931.5 MeV
electron mass (M_e)	= 0.5110 MeV
mass of a hydrogen atom (M_H)	= 938.789 MeV
mass of a neutron (M_n)	= 939.572 MeV

According to the proton-neutron hypothesis, the number of protons in a nucleus equals its positive charge and hence its *atomic number Z*. The total number of nucleons equals its *mass number A*. Consequently, the *neutron number N* is A–Z (A, Z and N are integers).
Knowing Z, N, M_H and M_n we can calculate the mass of all the components of an atomic nucleus. Comparing this with the atomic mass M (determined by mass spectrometry) we obtain a mass difference which is called the *mass defect* ΔM:

$$Z M_H + N M_n - M = \Delta M \tag{2}$$

The mass defect, which can also be expressed in units of energy according to *Einstein*'s Formula, equals the *binding energy of the nucleus B* and is the amount of energy necessary to cause a nucleus to disintegrate into protons and neutrons.

$\frac{B}{A}$ is the average binding energy per nucleon.

Figure 1.1. shows that light and heavy nuclei show comparatively small values of $\frac{B}{A}$, whereas nuclei of medium masses have higher average nucleon binding energies, indicating their greater stability.

According to basic laws of thermodynamics it should be possible to generate energy by the splitting of heavy nuclei or by the fusion of light ones. Both

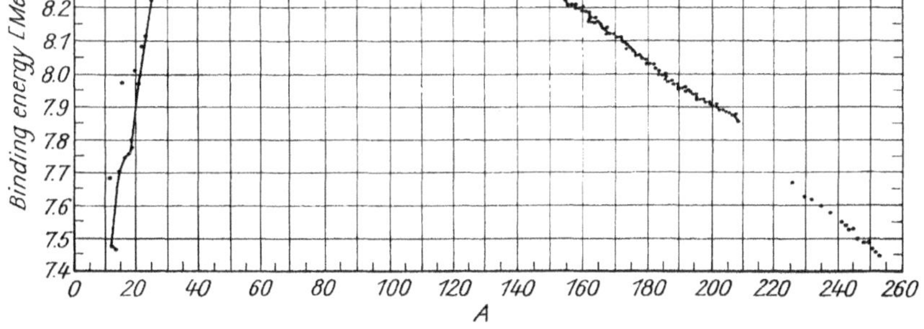

Fig. 1.1. $\frac{B}{A}$ as a function of A

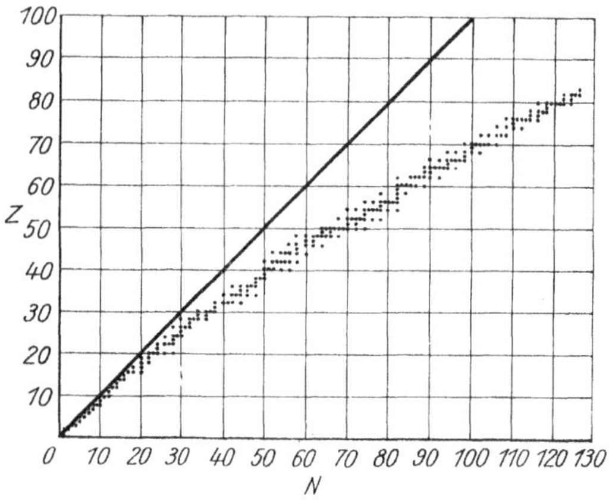

Fig. 1.2. Plot of Z versus N

these processes have been technically realized in nuclear reactors and in hydrogen bombs respectively. Moreover, the sun's radiant energy results from the fusion of hydrogen atoms to form helium atoms.

2. Nuclear systematics

According to the proton-neutron model the identity of a nucleus, in a sense, is given by the number of protons and neutrons. One can obtain rather simple systematics by plotting the known stable and radioactive nuclides in a graph as given in Figure 1.2. This is called the chart of nuclides. Nuclides with the same number of protons (and therefore belonging to the same chemical element) but with different numbers of neutrons are called *isotopes.* They are plotted on the horizontal axis. Vertically the nuclides are arranged in the order of increasing atomic (or proton) number Z.
Nuclides with the same mass number A, the so-called *isobars* lie on the diagonal axis from top left to bottom right. Furthermore Figure 1.2. shows that the light, stable nuclei ($Z \leq 40$) exhibit a neutron-proton ratio of unity. This ratio gradually increases with growing A, indicating an excess of neutrons. This is also called the isotope number.

3. Rules for radioactive transformation

3.1. Alpha decay

Alpha decay occurs mainly with nuclides of $Z > 83$ (Bi). Their instability results from the Coulombic repulsive force between protons. In the chart of nuclides, an alpha decay (the emission of two protons and two neutrons) is symbolized by a step extending over 4 mass units (A – 4) parallel to the N = Z diagonal. Consequently it leads to nuclei with an excess of neutrons.

3.2. Beta transformation

From Figure 1.2. we have learned that for each mass number the stability of a nucleus is characterized by a particular proton-neutron ratio. Otherwise the nuclei have a deficiency or a surplus of neutrons and undergo beta transformation.

According to

$$n \rightleftharpoons p + \beta + E + \nu \qquad (3)$$

an isolated neutron transforms spontaneously (half-life 10 m) into a proton with energy release (0.78 MeV), emitting a beta particle (i.e. an electron) and a neutrino ν. The neutrino is an almost massless uncharged particle. Equation (3) is valid also for excess neutrons in an atomic nucleus, however, with an energy release and half-life characteristic of the nuclide in question. Reaction (3) is reversible and hence in principle also valid for nuclides with a surplus of protons (i.e. deficiency of neutrons).

Beta transformations are accompanied by the emission of electrons. The mass number of the transforming nucleus remains unchanged. Parent and daughter nuclide are isobars.

To explain some of the features of beta transformation we shall use the *Weizsäcker* formula which is based on a semiempirical approach relating the binding energy (or total mass) of any nucleus to its composition (Z and A):

$$B = a_v A - a_O A^{2/3} - \frac{a_c Z^2}{A^{1/3}} - a_a \frac{(A-2Z)^2}{A} \pm \delta A^{-1} \qquad (4)$$

The first term expresses the fact that the binding energy is nearly proportional to A.

The second term takes into account the fact that at the surface of a nucleus the nucleons have unsaturated forces and consequently reduced binding energy proportional to the surface area ($A^{2/3}$).

The third term takes into consideration the Coulombic repulsive force between protons in a nucleus. This term accounts for the fact that all stable nuclei with $Z > 20$ contain more neutrons than protons.

The fourth term is based on the assumption that nuclei tend to have equal numbers of neutrons and protons. This is indeed true of the light nuclei ($A \leq 40$). The deviation from this trend among heavier nuclei can be accounted for by the Coulomb effect. The magnitude of this term increases with increasing value of (A – 2 Z), the neutron excess.

The final term expresses the fact that the binding energy for a given A depends on wether N and Z are even or odd.

So-called even-even nuclei are the most stable. For them δ is positive, for odd-even and even-odd nuclei $\delta = 0$ and for odd-odd nuclei δ is negative.

The combination of equations (2) and (4) assuming M = B results in a formula in which the atomic mass M is represented as a function of A and Z.

$$M = ZM_H + (A-Z)M_n - a_V A + a_O A^{2/3} + \frac{a_c Z^2}{A^{1/3}} + a_a \frac{(A-2\,Z)^2}{A} \pm \sigma A^{-1} \qquad (5)$$

For constant A this is a quadratic equation in M and Z and therefore represents a parabola for a given value of σ.
For an odd A a single parabola is the result (Figure 1.3.), while a nucleus with an even A is represented by two parabolas, separated from each other by $2\,\sigma/A$ on the energy axis (Figure 1.4.).
In other words, the binding energies of all nuclei can be represented as a function of A and Z by means of a three-dimensional plot of equation (4). A vertical section through this nuclear energy surface results in single or

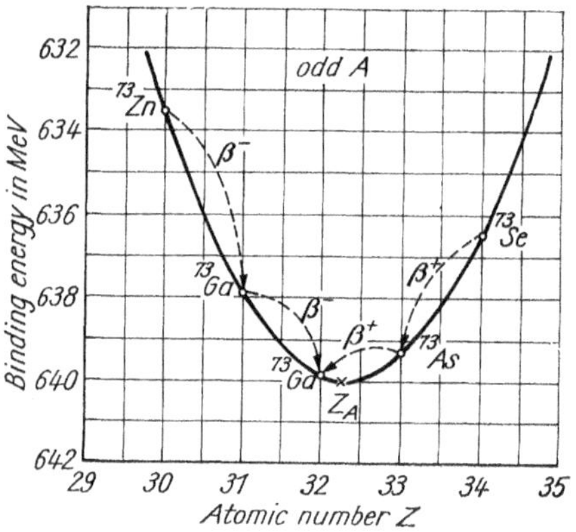

Fig. 1.3. Binding energy of nuclides with A = 64

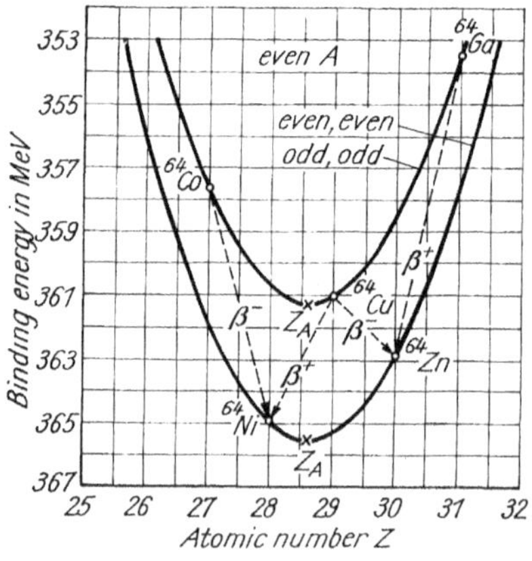

Fig. 1.4. Binding energy of nuclides with A = 73

double parabolas. The connecting line between their vertices forms the so-called "energy valley" the bottom of which, is occupied by the stable nuclides, while the beta unstable nuclides are placed on its two slopes.

4. Radionuclides, half-life, specific activity

In contrast to stable isotopes, radionuclides disintegrate spontaneously at a velocity characteristic of the nuclide involved, and emit nuclear radiation. In this process the disintegrating parent nuclide which has a higher energy level, is transformed into the daughter nuclide which represents a lower energy level. The daughter nuclide, again, may be radioactive. The emitted nuclear radiation can easily be measured, as a rule, by means of a suitable detector, e.g. a GM-counter or a scintillation counter. From this, follows the possibility of detecting nuclides in very small quantities, especially those which disintegrate rapidly.
The quantity of a radionuclide is not normally quoted in units of mass (g, mg, μg or ng)—although this is possible in principle—(see below), but is quoted in units of activity.

Table 1.1. The half-life and the disintegration constant for some short-, medium-, and long-lived radionuclides

Radionuclide	Half-life	$t_{1/2}$	λ [s^{-1}]
^{11}C	20.4	m	$8.17 \cdot 10^{-4}$
^{24}Na	15	h	$1.85 \cdot 10^{-5}$
^{99m}Tc	6	h	$4.63 \cdot 10^{-5}$
^{125}I	60	d	$1.93 \cdot 10^{-7}$
^{170}Tm	120	d	$9.65 \cdot 10^{-8}$
^{192}Ir	74	d	$1.56 \cdot 10^{-7}$
^{14}C	5,700	a	$5.56 \cdot 10^{-12}$
^{238}U	$4.51 \cdot 10^{9}$	a	$7.03 \cdot 10^{-18}$
^{239}Pu	$2.44 \cdot 10^{4}$	a	$1.3 \cdot 10^{-12}$

m — minute; h — hour; d — day; a — year

5. The unit of the radioactivity is the bequerel (Bq)

Which is the quantity of a radionuclide suffering one radioactive disintegration in one second. (Formerly the unit 1 Ci $= 3.7 \cdot 10^{10}$ disintegrations per second was used.) For practical purposes it is sometimes advisable to use multiples of the unit Bq:

1 kBq (kilobequerel) = 1,000 dps (disintegrations per second)
1 MBq (megabequerel) = 10^6 dps
1 GBq (gigabequerel) = 10^9 dps
The rate of the radioactive disintegrations is usually described as the half-life: that is, the time interval in which 50 % of the activity of a radionuclide has decayed (Table 1.1.).

Mathematically this can be expressed as follows:

$$-\frac{dN}{dt} = N\lambda \qquad (6)$$

$-\frac{dN}{dt}$ is the rate of disintegration of a radionuclide

N is the number of radioactive nuclei
λ is the disintegration constant which is characteristic of a radionuclide

In other words, the rate of the disintegration is proportional to the quantity of a radionuclide present at any moment. In the course of time, the number of nuclei disintegrating per unit time decreases, according to the following equation (which is obtained by means of a mathematical transformation)

$$N = N_0 e^{-\lambda t} \qquad (7)$$

with N_0 indicating the number of radioactive nuclei existing at the time 0, and t the time that has elapsed from the time t = 0. By means of this equation the quantity (activity) of a radionuclide left over after a certain time can be calculated. It is easier, however, to use the following curves (Figure 1.5.) which were calculated by means of equation (7).
The half-life $t_{1/2}$ is connected with the disintegration constant by the following relationship:

$$t_{1/2} = \frac{\ln 2}{\lambda} = \frac{0.69315}{\lambda} \; (t_{1/2} \text{ in seconds!}) \qquad (8)$$

In practice it is often necessary to know the *mass activity concentration S* (formerly specific activity) of the radionuclide used.
The mass-activity concentration S is the quotient of the activity and mass of a radionuclide. S can be calculated for each radionuclide from its half-life.

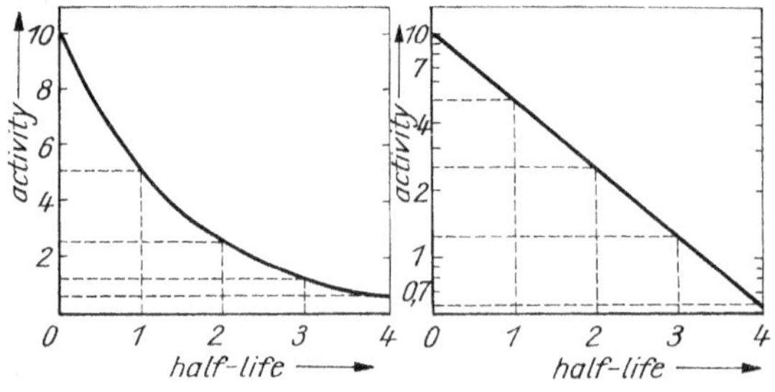

Fig. 1.5. Decay curves with half-life plotted in linear and logarithmic scale

From equations (6) and (8) it follows that

$$-\frac{dN}{dt} = \frac{N \cdot 0.693}{t_{½}}$$

If we substitute into this equation

$$-\frac{dN}{dt} = C\ [\mathrm{Bq}] \text{ and}$$

$$N = \frac{m \cdot L}{M \cdot t_{½}}$$

L = Avogadro's number = $6.02 \cdot 10^{23}\ \mathrm{mol}^{-1}$
m = mass of the nuclide [g]
M = atomic mass [a.m.u.]
$t_{1/2}$ = half-life [s]

We get

$$\frac{C}{m} = S = \frac{L \cdot 0.69315}{M \cdot t_{½}} = \frac{4.18 \cdot 10^{23}}{M \cdot t_{½}} \left[\frac{-\mathrm{Bq}}{\mathrm{g}}\right] = \frac{4.18 \cdot 10^{17}}{M \cdot t_{½}} \left[\frac{-\mathrm{MBq}}{\mathrm{g}}\right]$$

(half-life in s)

$$= \frac{4.18 \cdot 10^{17}}{M \cdot 0.864 \cdot 10^{5} \cdot t_{½}} = \frac{4.838 \cdot 10^{12}}{M \cdot t_{½}} \left[\frac{-\mathrm{MBq}}{\mathrm{g}}\right]$$

(half-life in d)

$$= \frac{4.18 \cdot 10^{17}}{M \cdot t_{½} \cdot 3.1558 \cdot 10^{7}} = \frac{1.325 \cdot 10^{10}}{M \cdot t_{½}} \left[\frac{-\mathrm{MBq}}{\mathrm{g}}\right]$$

(half-life in a)

For the radionuclides listed in Table 1.1., the following values of the specific activity are obtained (Table 1.2.).

Table 1.2. Mass activity concentrations

Nuclide	$S\left[\frac{\mathrm{MBq}}{\mathrm{g}}\right]$	$S\left[\frac{\mathrm{Ci}}{\mathrm{g}}\right]$
^{11}C	$3.1 \cdot 10^{13}$	$8.38 \cdot 10^{8}$
^{24}Na	$3.23 \cdot 10^{11}$	$8.73 \cdot 10^{6}$
^{99m}Tc	$1.95 \cdot 10^{11}$	$2.57 \cdot 10^{6}$
^{125}I	$6.45 \cdot 10^{8}$	$1.73 \cdot 10^{4}$
^{170}Tm	$2.37 \cdot 10^{8}$	$6.4 \cdot 10^{3}$
^{192}Ir	$3.41 \cdot 10^{8}$	$9.2 \cdot 10^{3}$
^{14}C	$1.66 \cdot 10^{5}$	4.49
^{238}U	$1.23 \cdot 10^{-2}$	$3.32 \cdot 10^{-7}$
^{239}Pu	$2.27 \cdot 10^{3}$	$6.14 \cdot 10^{-2}$

6. Modes of radioactive disintegration

6.1. Alpha decay

Provided a heavy nucleus is stable as regards beta transformation (i.e. it lies in the "energy valley") it can reduce its surplus in positive charge only by emission of alpha particles (i.e. helium ions of mass 4 and charge + 2). The transformation from a defined energy level of the parent nuclide can lead either to the ground level of the daughter nuclide and/or to excited state of the daughter. Consequently, alpha emitters exhibit a mono- or polychromatic spectrum.

The de-exitation of an excited state occurs through the emission of gamma radiation. This occurs very often with nuclei which have an odd atomic number.

Alpha emitters exhibit a discrete spectrum in respect to alpha and gamma radiation. This is in contrast to beta emitters.

6.2. Beta decay

Equation (3) describes the transformation of an isolated neutron. Likewise, a surplus neutron in a given atomic nucleus can transform, e.g.

$$^{32}_{15}P \rightarrow {}^{32}_{16}S + \beta^- + \nu, \qquad (10)$$

in this case, however, with a half-life of 14 d characteristic of ^{32}P and the disintegration energy of $E_{\beta\,max} = 1.7$ MeV.

Another example is the β-disintegration of tritium, a heavy isotope of hydrogen,

$$^{3}_{1}H \rightarrow {}^{3}_{2}He + \beta^- + \nu, \qquad (11)$$

The corresponding disintegration data are:

half life = 12.3 a

disintegration energy: $E_{\beta\,max}$ = 0.0186 MeV

Tritium is a nuclide often used as a tracer in organic and biochemistry.

Incidentally, equations (10) and (11) show that in β^--disintegration the mass number (= number of nucleons) is maintained but that the atomic number (proton number) of the nucleus is increased by one. The particles emitted by beta emitters show a continous energy distribution (beta-ray spectrum) in contrast to the alpha particles or gamma photons produced by nuclear disintegration. The upper limiting value of the energy of the β-particles ($E_{\beta\,max}$) characteristic of a particular nuclide is given in all nuclide tables. It should be kept in mind that the actual energy of the beta particles is always below this limiting value. Figure 1.6. shows a typical beta-ray spectrum.

A rule of thumb which can be applied is that, *on the average,* a beta-particle has an energy of $^1/_3\,E_{\beta\,max}$.

In contrast to β^--disintegration, a neutron deficiency of the nucleus leads to what is called *electron capture (E.C.) or to β^+-emission.*

Since the radioactive atoms are not observed directly but only by their radioactive decay it is reasonable to substitute A and A_0 for N and N_0, respectively, representing the *activity* at t and t = 0.

Assuming $\frac{A}{A_0} = \frac{1}{2}$ (i.e. the activity has decreased to $\frac{1}{2}$ of A_0 during a characteristic time $t_{1/2}$ = half-life) one can calculate:

$$\frac{1}{2} = e^{-\lambda t_{1/2}}$$

$$2 = e^{\lambda t_{1/2}}$$

$$\ln 2 = e^{\lambda t_{1/2}} = 0.693$$

$$t_{1/2} = \frac{0.693}{\lambda} (2) = (\text{eq. } 8 \text{ of } 5.)$$

The average life expectancy can be calculated as the sum of the time of existence of all the atoms of a radioactive species under consideration divided by N_0:

$$\text{Average life } \tau = \frac{1}{\lambda} \tag{3}$$

τ is the time interval during which an amount of activity decays to $\frac{1}{e}$ of its initial value.

Growth of radioactive products

The daughter nuclide resulting from the disintegration of a parent nuclide is very often radioactive itself. Hence the decay rate of the parent determines the growth of the daughter. We denote the parent nuclide by index 1 and the daughter nuclide by index 2 (hereafter we use N_1^0 instead of N_0 to denote the parent atoms at t = 0).

From (1) and (2) we obtain:

$$\frac{dN_2}{dt} = \lambda_1 N_1 - \lambda_2 N_2 \tag{4}$$

The growth of the daughter $\frac{dN_2}{t}$ equals the decay of the parent $\lambda_1 N_1$ diminished by the decay of the daughter $\lambda_2 N_2$. The transformation of (4) qields:

$$\frac{dN_2}{dt} + \lambda_2 N_2 - \lambda_1 N_1^0 e^{-\lambda_1 t} = 0$$

Solving this linear differential equation gives:

$$N_2 = \frac{\lambda_1}{\lambda_2 - \lambda_1} N_1^0 (e^{-\lambda_1 t} - e^{-\lambda_2 t}) + N_2^0 e^{-\lambda_2 t} \tag{5}$$

N_2^0 is the value of N_2 at t = 0

Or, with respect to activity:

$$A_2 = A_1^0 \frac{\lambda_2}{\lambda_2 - \lambda_1} (e^{-\lambda_1 t} - e^{-\lambda_2 t}) + A_2^0 e^{-\lambda_2 t} \tag{6}$$

A_1^0 and A_2^0 are the activities at $t = 0$ of the parent and daughter, respectively. The term in brackets shows the growth of the daughter from the parent $e^{-\lambda_1 t}$ and its decay $e^{-\lambda_2 t}$ while the last term represents the daughter atoms present initially.

Another approach leads to a special case of equation (5): In the process of radioactive transformation only

$$N_1 = N_1^0 e^{-\lambda_1 t}$$

of the initially present radioactive atoms N_1^0 have survived the time interval t, the other part

$$N_1^0 - N_1 = N_1^0 (1 - e^{-\lambda t}) \qquad (7)$$

has decayed by forming daughter atoms. So we can write

$$N_1^0 - N_1 = N_2.$$

From eq. (5) we obtain eq. (7) by setting

$\lambda_2 = 0$ since decay of daughter atoms is not considered

and

$N_2^0 = 0$ since we assume that no daughter is initially present

$$N_2 = -N_1^0 (e^{-\lambda_1 t} - 1)$$

$$N_2 = N_1^0 (1 - e^{-\lambda_1 t}) \qquad (8)$$

eq. (9) is important in radiometric dating and in a similar form in neutron activation.

Radioactive equilibrium

On the basis of eq. (5) we can distinguish two general cases, namely, whether the parent or the daughter has the longer half-life. In the first case ($\lambda_1 < \lambda_2$) radioactive equilibrium can be attained. That is, the ratio of the disintegration rates of parent and daughter becomes constant.

From (5) we can see that after t becomes sufficiently large $e^{-\lambda_2 t}$ becomes negligible.

$$N_2 = \frac{\lambda_1}{\lambda_2 - \lambda_1} N_1^0 e^{-\lambda_1 t} \quad \text{since}$$

$$N_1^0 e^{-\lambda_1 t} = N_1$$

$$\frac{N_1}{N_2} = \frac{\lambda_2 - \lambda_1}{2} \qquad (9)$$

Substituting into (6) the activities of parent and daughter:

$$A_1 = \lambda_1 N_1$$
$$A_2 = \lambda_2 N_2$$

we obtain:

$$\frac{A_1}{A_2} = \frac{\lambda_2 - \lambda_1}{\lambda_2} = 1 - \frac{\lambda_1}{\lambda_2} \qquad (10)$$

(It has to be emphasized that the energy of radiation (or the character) of parent and daughter nuclides are not identical and hence they are detected with different efficiences.)
Consequently, the measured activity has to be multiplied by the efficiency factor to attain the theoretical activity which we have assumed in our consideration so far.
The ratio of the activities has values between 0 and 1 depending on λ_1 and λ_2. In equilibrium, the daughter activity is always higher than that of the parent by the factor $\frac{\lambda_2}{\lambda_2 - \lambda_1}$ and both activities decay with the half-life of the parent. This case is the so-called *transient equilibrium* and is illustrated in Figure 1.7.
A limiting case of equilibrium in which $\lambda_1 \ll \lambda_2$ is the *secular equilibrium.* The parent activity does not decrease measurably during many daughter half-lifes.

Instead of (6) we use the approximation

$$\frac{N_1}{N_2} = \frac{\lambda_2}{\lambda_1} \text{ or } N_1\lambda_1 = N_2\lambda_2 \tag{11}$$

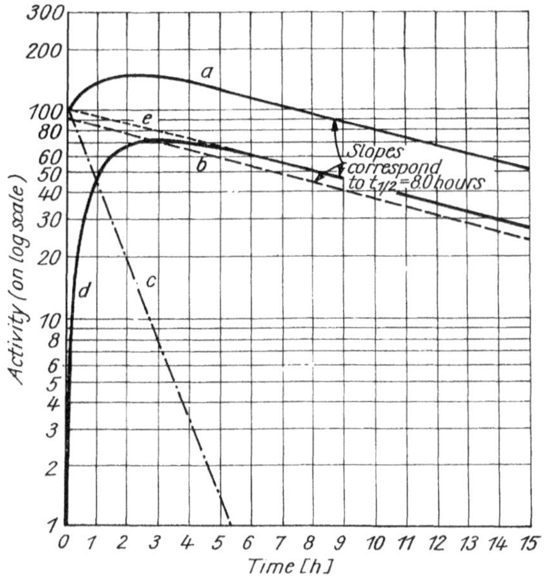

Fig. 1.7. Transient equilibrium
a = total activity b = parent activity c = decay of isolated daughter d = daughter growing in purified parent fraction e = total daughter in parent-plus-daughter fraction

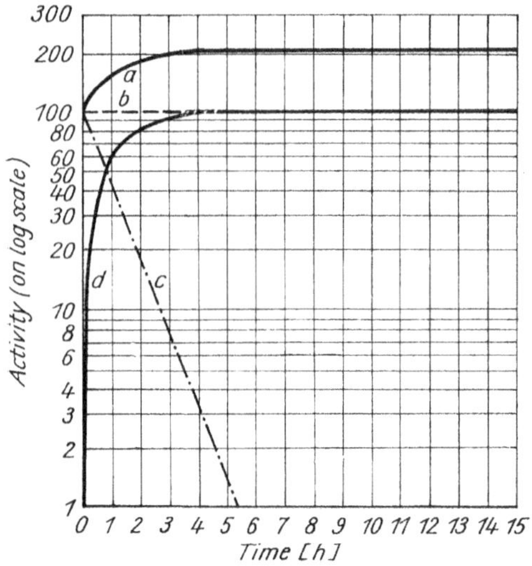

Fig. 1.8. Secular equilibrium
a = total activity b = parent activity c = decay of isolated daughter d = daughter in purified parent fraction

The parent and daughter activities are equal:

$$A_1 = A_2$$

The secular equilibrium is illustrated in Figure 1.8.

No equilibrium

In the case of the parent being shorter-lived than the daughter ($\lambda_1 > \lambda_2$) no equilibrium is attained at any time.

Assuming we start with a parent free of daughter nuclide, the latter rises as the parent decays. The daughter activity passes a maximum at the time t_m and eventually decays with its characteristic half-life, Figure 1.9.

To find t_m we differentiate eq. (5):

$$\frac{dN_2}{dt} = -\frac{\lambda_1^2}{\lambda_2 - \lambda_1} N_1^0 e^{-\lambda_1 t} + \frac{\lambda_1 \lambda_2}{\lambda_2 - \lambda_1} N_1^0 e^{-\lambda_2 t}$$

and set

$$\frac{dN_2}{dt} = 0 \text{ if } t = t_m.$$

$$\frac{\lambda_2}{\lambda_1} = e^{(\lambda_2 - \lambda_1)\, t_m}$$

or

$$t_m = \frac{2.303}{\lambda_2 - \lambda_1} \log \frac{\lambda_2}{\lambda_1} \qquad (12)$$

If the decay curve of the daughter is extrapolated back to $t = 0$ the intercept measures the apparent activity $N_1^0 \lambda_2$.

Bateman equation

For the general case of many successive decays *Bateman* has given a solution of the differential equation of a chain of n members. Assumed at $t = 0$ only the parent nuclide is present i.e.

$$N_2^0 = N_3^0 = N_n^0 = 0,$$

then follows:

$$N_n = C_1\, e^{-\lambda_1 t} + C_2\, e^{-\lambda_2 t} + \ldots C_n\, e^{-\lambda_n t} \tag{13}$$

$$C_1 = \frac{\lambda_1 \lambda_2 \ldots \lambda_{n-1}}{(\lambda_2 - \lambda_1)(\lambda_3 - \lambda_1)\ldots(\lambda_n - \lambda_1)} N_1^0$$

$$C_2 = \frac{\lambda_1 \lambda_2 \ldots \lambda_{n-1}}{(\lambda_1 - \lambda_2)(\lambda_3 - \lambda_2)\ldots(\lambda_n - \lambda_2)} N_1^0$$

Branching decay

Sometimes a parent nuclide may decay in two different ways forming different daughters:

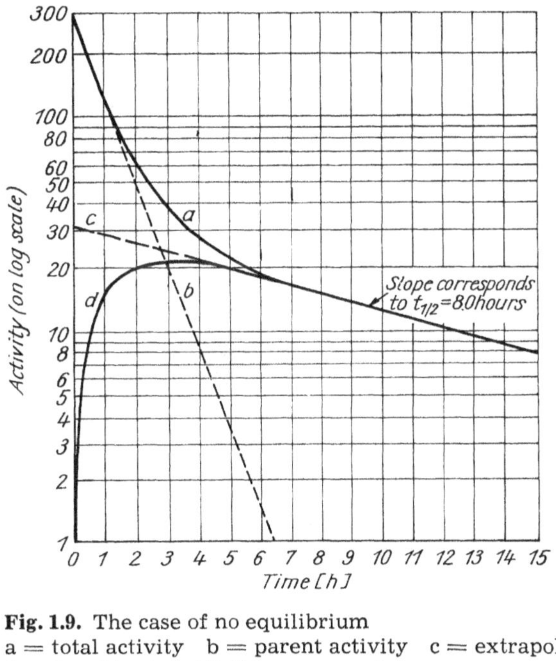

Fig. 1.9. The case of no equilibrium
a = total activity b = parent activity c = extrapolation of final decay to $t = 0$
d = daughter in initially pure parent

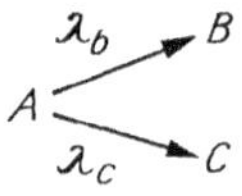

We have to consider two decay constants λ_b and λ_c, but A can decay only with one half-life:

$$t_{1/2} = \frac{0.693}{(\lambda_b + \lambda_c)}$$

B and C are formed according to $N_A\ \lambda_b$ and $N_A\ \lambda_c$ respectively.

References

1. *Friedlander, G., J. W. Kennedy, E. S. Macias, J. M. Miller:* Nuclear and Radiochemistry. John Wiley and Sons, New York, London, Sydney 1981.
2. *Choppin, G. R., J. Rydberg:* Nuclear Chemistry. Pergamon Press, Oxford, New York, Toronto, Sydney, Frankfurt 1980.
3. *Lieser, K. H.:* Einführung in die Kernchemie. Verlag Chemie, Weinheim/Bergstr. 1980.

Chapter 2

Modern Trends in Isotope Production with Small Accelerators

Gerd-Juergen Beyer, Rudolf Muenze

1. General aspects

Short-lived cyclotron isotopes do not emit β^--particle in contrast to the overwhelming part of reactor-produced isotopes; they are carrier-free and show mostly favourable gamma-radiation characteristics. The β^+-radiation of the "biological" isotopes ^{11}C, ^{13}N, ^{15}O and ^{18}F allows the coincidence technique for registration with a higher resolution in dynamic tomographic studies to be employed.

These are the main reasons for the need of cyclotron isotopes, which has rapidly increased.

More than 60 cyclotrons all over the world are installed to produce medically useful radioisotopes with different rates of utilization. A short review of characteristics of the cyclotron – types and main products will be given in the following.

Cyclotron-produced isotopes were introduced into the biological and medical fields as early as the thirties. The first cyclotron completely applied to research and application in medicine was installed in 1955/56 at Hammersmith Hospital, London; the next one entirely used in the medical field came into operation nearly 10 years later at Washington University in 1965. But in the early seventies a quite large number of cyclotrons were installed especially at medical centres. They were designed to produce new radiopharmaceuticals. About six companies have designed a number of special cyclotrons suitable for application purposes. But in the late seventies these companies were faced with increasing difficulties to sell the commercial "compact"-cyclotrons.

1.1. Characterization of accelerator-produced isotopes

Radioisotopes produced by charged particle reactions at small electrostatic medium energy—or high energy accelerators differ considerably from those radioisotopes, produced normally in nuclear reactors through (n, γ) or fission reaction. Accelerator-produced isotopes are normally neutron-deficient nuclides, which do not emit β^--particles; normally they decay by EC- and mostly in cases of short half-lives by emitting β^+-particles. Very often such isotopes show better gamma-ray characteristics with regard to nuclear medical application than most reactor-produced isotopes.

Accelerator-produced isotopes are mostly carrier-free, while the often mentioned short half-life is not a specific quality of cyclotron isotopes.

There are basically two reasons leading to an increasing need for cyclotron-produced isotopes and therefore to a need for the relevant "service-friendly" compact cyclotron:

1. short-lived cyclotron-produced radioisotopes are more favourable than the other reactor-produced isotopes (except ^{99m}Tc), because of the absence of β^--emission. The change to neutron-deficient isotopes decreases the radiation dose to patient and staff. Due to better gamma-ray properties high quality imagings may be achieved. The carrier-free quality leads to a simplification of the labelling procedure. Short half-lives in connection with low doses allow repeat imaging procedures in short intervals. By using the coincidence technique for registration of β^+-emitting short-lived isotopes, high quality imagings can be obtained.
2. More important, however, seems to be the change in the character of radioisotope application in nuclear medicine. While in studies of organ morphology radiopharmaceuticals are increasingly replaced by other modern methods as ultrasonics or computer tomography, the availability of a number of cyclotron isotopes and compounds labelled with these nuclides and their direct involvment in functional and metabolic processes opens a whole new range of applications in "non-invasive" dynamic measurements of physiological and biochemical processes in vivo. Developments in this direction are irreparably connected with cyclotron-produced isotopes, especially such as ^{11}C, ^{13}N, ^{15}O and ^{18}F.

Due to these excellent advantages and possibilities one should hope for a growing spread of cyclotron-isotopes in the field of nuclear medicine, but there are several reasons for the delay of such developments.

1. As far as modern developments in the field of radiopharmacology are concerned, the development of detection and registration methods convenient for short-lived cyclotron-produced isotopes were delayed. This disproportion seems to be overcome, since at present remarkable activities can be observed.
2. Cyclotrons including the small "compact cyclotrons" are valuable complicated facilities, whose effective operation necessarily needs a corresponding technological basis as well, such as equipped workshop, hot laboratories a trained multidisciplinary technical staff for maintenance and a reliable distribution system.

Thus *Homer B. Hupf* from the Mont Sinai Medical Centre in Miami Beach, Florida, mentioned already in 1973 after installations of a Cs-30-cyclotron: "The main shortcoming in the compact cyclotron is the lack of target systems capable of removing the head deposited in the target material. An institution contemplating the installation of a compact cyclotron should plan for a target development program."
Meeting these difficulties, the IAEA founded a research coordination programme on "Production of Radiopharmaceuticals from Accelerator-Produced Isotopes": The Central Institute for Nuclear Research of the GDR-Academy of Sciences at Rossendorf is involved in this research programme.

1.2. Present state of cyclotron application for isotope production

Today more than 60 cyclotrons are in operation all over the world partly or entirely used for radioisotope production, mainly for nuclear medical application. These machines may be divided into two groups: commercially and noncommercially used cyclotrons (Table 2.1. and Table 2.2.). Both groups do not differ in the type of accelerator but they evidently differ in the effectivity of utilization and in product assortment. The cyclotrons, operated by the large isotope-producer companies as NEN or Amersham, are distinguished by a well developed remote-controlled target irradiation equipment. These companies run two or even three cyclotrons, which are being operated nearly automatically and utilized with about 100 % efficiency. The products are normally gamma-emitting isotopes for medical application both with half-lives of a few days, such as ^{67}Ga, ^{111}In and ^{201}Tl, and long-lived isotopes for application in medicine and other fields, such as ^{57}Co, ^{195}Au, ^{109}Cd, and ^{44}Ti. Of the shorter-lived radioisotopes only ^{81}Rb and ^{123}I are introduced into the production programme in limited amounts, whereas the ultra-short-lived β^+-emitting isotopes are not essentially synthesized. This production profile already points to the problem of distributing modern cyclotron-produced radiopharmaceuticals. The cyclotrons installed at nuclear medical centres are utilized for 20–40 hours per week only for radionuclide production.
Another part of beam time is used for neutron therapy at some centres. Normally the ultra-short-lived β^+-isotopes are produced, whereas longer-lived gamma-emitting isotopes, such as ^{67}Ga or ^{111}In, are produced only at a few centres. Priority is given to the radioisotopes ^{81}Rb and ^{123}I in nearly

Table 2.1. Cyclotrons used for commercial large-scale isotope production (1980)

Company	Type of cyclotron	Main products
USA		
Medi Physics	Scanditronix M 40	^{81}Rb, ^{123}I, ^{201}Tl
	Cs-22	^{67}Ga, ^{81}Rb, ^{111}In, ^{123}I, ^{201}Tl
	Cs-22	^{67}Ga, ^{81}Rb, ^{123}I
New England	Cs-22	^{57}Co, ^{67}Ga, ^{109}C ^{111}In
Nuclear	Cs-30	^{68}Ge, ^{195}Au, ^{201}Tl
	Cs-30	^{68}Ge, ^{195}Au, ^{201}Tl
UK	Philips	^{67}Ga, ^{111}In, ^{57}Co, ^{73}As
Amersham	(27 MeV p)	^{55}Fe, ^{52}Mn, ^{44}Ti, ^{48}V
The Netherlands	Philips	
Mallinckrodt	(28 MeV p)	no information
Soviet Union	NIIEFA	^{67}Ga, ^{57}Co
Obninsk	U-150	long-lived isotopes for application
Kentor IZOTOP	(22 MeV p)	in other than medical fields
Japan	Cs-30	^{18}F, ^{67}Ga, ^{81}Rb, ^{111}In, ^{123}I, ^{201}Tl

Table 2.2. Cyclotrons at medical or nuclear research centres completely or partly designed for isotope production (1980)

Institution	Type of cyclotron	Beam-time for isotope production	Main isotope
USA			
Michigan	VEC 50 MeV p	6 h/w	^{11}C, ^{13}H
New York	Cs 15	20 h/w	^{11}C, ^{13}H, ^{15}O, ^{18}F, ^{38}K, ^{52}Fe
Oak Ridge	Isochron 55 MeV p	2 %	^{67}Ga and others
St. Louis	Allis Chalmers	?	^{11}C, ^{13}N, ^{15}O, ^{18}F
	Cs 15	20 h/w	^{11}C, ^{13}N, ^{15}O, ^{18}F
	52 inch (30 MeV)	12 h/w	^{77}Br, ^{77}Kr, ^{81}Rb
Washington	65 MeV p	?	^{11}C, ^{13}N, ^{18}F
Boston	Allis Chalmers	40 h/w	^{11}C, ^{13}N, ^{18}F, ^{15}O
Bloomington	VEC 200 MeV p	1 %	^{123}I
Brookhaven	60 inch (36 MeV p)	40–50 h/w	^{11}C, ^{13}N, ^{15}O, ^{18}F, ^{38}K, ^{81}Rb, ^{123}I, ^{211}At
	BLIP	145 h/w	^{28}Mg, ^{52}Fe, ^{67}Ga, ^{81}Rb, ^{97}Ru, ^{123}I
Chicago	Cs 15	10 h/w	^{11}C, ^{13}N
Cleveland	VEC (45 MeV p)	8 h/w	^{11}C, ^{201}Tl
Davis	VEC (66 MeV p)	20–35 h/w	^{123}I, ^{201}Tl
Los Alamos	LAMPF (800 MeV p)	168 h/w	^{52}Mn, ^{52}Fe, ^{77}Se, ^{77}Br, $^{82}Sr/Rb$, $^{172}Hf/lu$
Los Angeles	Cs 22	50 h/w	^{11}C, ^{13}N, ^{18}F, ^{62}Zn, ^{62}Cu, ^{81}Rb
Miami	Cs 30	?	^{11}C, ^{15}O, ^{67}Ga, ^{111}In, ^{201}Tl
The Netherlands			
Eindhoven	AVF (30 MeV p)	8 h/w	^{123}I, ^{77}Br
Groningen	AVF (65 MeV p)	7 h/w	^{11}C, ^{13}N, ^{55}Co
Switzerland			
Würenlingen	VEC (72 MeV p)	3 h/w	^{123}I, ^{111}In
UK			
Birmingham	10 MeV p	20 h/w	^{81}Rb, ^{123}I, daily
Edinburgh	Cs 30		^{81}Rb
Harwell	VEC (58 MeV p)	12 h/w	^{123}I
Hammersmith	45 inch	37 h/w	^{11}C, ^{13}N, ^{15}O
London	(8 MeV p)		^{18}F, ^{81}Rb
	(30 MeV)		^{43}K, ^{52}Fe u. a.
Belgium			
Ghent	CGR-MeV 520	8 h/w	^{13}N, ^{81}Rb
Liege	CGR-MeV 520	25 h/w	^{11}C, ^{13}N, ^{18}F, ^{43}K, ^{81}Rb, ^{73}Se, ^{67}Ga, ^{111}In
Louvrin	CGR-MeV 930	15 h/w	^{52}Fe, ^{123}I
Finland			
Turku	MGZ (U-103)	16 h/w	^{11}C, ^{18}F, ^{81}Rb

Institution	Type of cyclotron	Beam-time for isotope production	Main isotope
France			
Orsay	CGR-MeV 520	30 h/w	^{11}C, ^{13}N, ^{15}O, ^{18}F, ^{43}K
FRG			
Jülich	CV-28	24 h/w	^{11}C, ^{18}F, ^{123}I
Essen	CV-28	50 %	^{11}C, ^{18}F, ^{123}I
Heidelberg	AEG		^{13}N, ^{18}F, ^{81}Rb, ^{123}I
Karlsruhe	26 MeV p	5 h/w	^{123}I
GDR			
ZfK Rossendorf	U-120 (7 MeV p) (28 MeV)	39 %	^{85}Sr, ^{67}Ga, ^{123}I, ^{81}Rb, ^{211}At

all cases. The leading position in this group of cyclotron owners is held by Hammersmith Hospital London.

From Table 2.3. one can see that the highest efficiency expressed in numbers of patients' investigations was achieved by the production of ^{81m}Kr-generators. For routine application the production of ultra-short-lived isotopes at this hospital has not yet played an important role. Because of the large extent of daily routine, there is not enough free capacity available which could be used for the development and introduction of short-lived metabolits. Indeed, the most favourable results were published by interdisciplinary groups working at nuclear research centres at Brookhaven, Orsay or Jülich.

Table 2.3. Routine clinical radionuclide production at Hammersmith Hospital (London) from January to September 1980

Nuclide	No. of production runs	Cyclotron time (hours)	Total activity	No. of investigated patients
^{11}C	3	4 : 56	on-line	4
^{13}N	11	3 : 33	on-line	19
^{15}O	68	273 : 46	on-line	150
^{18}F	24	19 : 17	199 mCi	48
^{81}Rb—^{81m}Kr	281	568 : 45	1,051 generators 33 mCi each	4,542
^{52}Fe	29	87 : 38	23.7 mCi	117
^{43}K	31	60 : 06	40.9 mCi	140
^{203}Pb	16	43 : 50	40.6 mCi	—
^{77}Br	38	135 : 45	144.7 mCi	186
	total	1,197.6 h (33 %)		5,206

1.3. Modern research subjects

According to the application of cyclotron-produced isotopes at present three major topics may be underlined:

1. Production and application of *^{81}Rb-^{81m}Kr-generators.* The ^{81m}Kr, available from such generators, is the shortest-lived radioisotope (13 s) at present used in nuclear medicine. It is applied as a gas in lung ventilation studies and as a solution in perfusion studies of the lung, heart and nearly all other organs. The ^{81m}Kr-191-keV-transition is very convenient for gamma-camera registration.
2. Application of *radioactive gases*
   Directly produced ^{11}C and ^{15}C are applied as gases in the form of ^{15}OO, ^{11}Co, $^{11}CO_2$, $C^{15}OO$ mainly in studies of heartblood circulation. Such investigations are often carried out in an online mode with the accelerator, and they require new modern detection and computing techniques, such as the positron emission transaxial tomographs, which are at present more expensive than the accelerator itself.
3. From the preliminary produced precursors with ^{11}C and ^{13}N, complicated metabolic compounds and pharmaceuticals may be already synthesized. Their application opens up completely new possibilities of studying pathological, physiological and metabolic processes *in vivo.*

2. The use of the Rossendorf U-120 cyclotron for isotope production

The general policy of our research work was

- to use accelerators of our institute for the production and application of radioisotopes being not available by reactor activation,
- to open up new fields for the application of radioisotopes, especially in medicine,
- to catch up with modern trends in radiopharmaceutical research.

The Rossendorf U-120 cyclotron is one of the oldest machines being still in operation. Nevertheless, despite the low proton-energy (7 MeV), the particle energy (d = 14 MeV, o = 28 MeV) allows us to keep abreast with modern developments in the field of nuclear medicine based on short-lived cyclotron-produced isotopes.

A modern remote-controlled target irradiation and handling facility was installed in 1977.

Since 1978, using the ^{nat}Zn (d, xn) ^{67}Ga-reaction, ^{67}Ga-citrate has been continuously produced.

A modern target system for irradiation of very sensitive materials (such as TeO_2) was developed. Using this system, ^{123}I may be produced by the reaction $^{122}TeO_2$ (d, n) ^{123}I in batches of $4 \cdot 10^9$ Bq (100 mCi). The 28 MeV o (-beam) is used for the production of ^{211}At and ^{81}Rb-^{81m}Kr-generators for lung ventilation studies.

At present 35–40 % of the cyclotron time available is utilized for isotope production. During the last few years this percentage has been kept nearly constant. But the percentage to produce the long-lived ^{85}Sr was displaced continuously. We then started the investigations for the production of the biological β^+-isotopes ^{11}C and ^{18}F. These isotopes may be produced with electrostatic accelerators, too.
During the last decade the interest of nuclear physicists in experiments with the obsolete U-120 cyclotron (25 years in operation) decreased considerably. The free time available was increasingly occupied by applied research being feasible with limited particle energies and intensities. Table 2.4. shows the beam-time distribution in 1980.
The major drawback of the machine is the very low proton energy insufficient for by far most production reactions published in the current literature. The very low fraction of proton activation results from these reasons. There are no reasonable approaches to changing this situation except the substitution of the accelerator itself. This will not be feasible in the next five years. The α-energies are quite suitable for the production of some useful radioisotopes, but until now the available intensity was too low for the production of radioisotopes in activities sufficient for application. In our opinion this problem can be solved by improving the α-particle production and the conditions of their acceleration. Some improvements have been already implemented. The distribution of the present beam-time is shown in Table 2.5.
From these figures, we estimate that there are further possibilities for the elaboration of more modern problems in accelerator-produced isotopes. The fairly large fraction of beam time spent on the production of ^{85}Sr is related to the utilization of free beam-time intervals occurring stochastically prior to or after nuclear physical experiments. In line with the results in target improvements, the production of this obsolete product is to be more and more replaced.
The time spent on development of new radiopharmaceuticals appears comparatively small. This is caused by the insufficient availability of more sophisticated target design in our group.
Thus, we recognized the development of appropriate target systems, meeting our conditions, as a major task in our work.

Table 2.4. Particle beam-time distribution at the Rossendorf U-120 cyclotron in 1980

Particle	Energy (MeV)	Fraction of total beam-time (4,834 hours/year) (%)
H_2^+	7	2
d	14	58
$^4He^{2+}$	28	29
$^3He^{2+}$	33	8

Table 2.5. User-spectrum of the Rossendorf U-120 cyclotron in 1980

Type of work	Beam-time (hours)	(%)
Nuclear physics	1,462	30.24
Neutron therapy	412	8.52
Isotope production	1,840	38.06
(^{67}Ga	547)	
(^{57}Sr	1,060)	
(isotope production development	233)	
Biophysics	242	5.01
Activation analyses	138	2.85
Surface activation for wearing and corrosion studies	193	4.00
Cyclotron improvement work	547	11.32
Total	4,834	
Shutdown maintenance	983	

2.1. Description of a remote-controlled system for target handling

Because of the very limited space at the experimental facilities of the Rossendorf U-120 cyclotron resulting in considerable disturbances of other operation by highly active irradiated targets, the construction of a remote-controlled target handling system became necessary.
A sketch of the design is shown in Figure 2.1. The system can be placed at any position of the beam line and allows a remote-controlled target transport to the irradiation position, the removal and introduction into a shielded container after irradiation, and the transport by means of a special lead-shielded chamber.
The handling system is placed directly behind the deflector electrode of the cyclotron. It consists of a transport carriage moving on rails and equipped with a shielded target chamber (1), a lift (2) for winding up the transport container (3), the target (6) and the target support (7). These parts are arranged on a strong steel frame (12). The irradiation position inside the vacuum line can be locked by means of a motor-driven vacuum valve. Reliable coupling of the vacuum valve and the target chamber by flanges (10) is achieved by a pneumatically operated cramp construction. The target holder consists of a water-cooled metal block (Figure 2.2.). The material to be irradiated is deposited onto the upper surface, predominantly by electroplating. Then the target holder is attached to the target support and locked.
The in- and outlet of the cooling line are fitted to the system through O-rings. Targets of different size corresponding to the special beam properties can be used. Furthermore, the target can be adjusted in arbitrary angular directions in order to regulate effective thickness and heat production. The total weight of the transport container is 23 kp, the target chamber is shielded by 10 cm of lead.

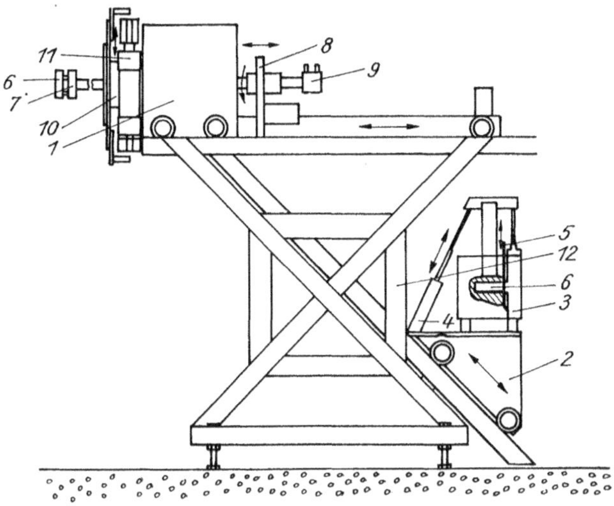

Fig. 2.1. Remote-controlled system for target irradiation and handling
1 = lead-shielded target chamber 2 = lift for the container 3 = container for target transport 4 = pneumatic cylinder for opening the container lid 5 = cramp 6 = target head 7 = target support 8 = target adjustment 9 = connection for water-cooling 10 = flange 11 = pneumatically operated cramp construction 12 = steel frame

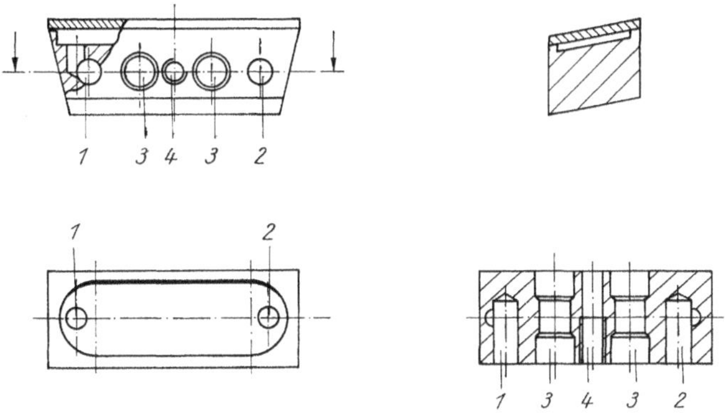

Fig. 2.2. Target head
1 = cooling water in 2 = cooling water out 3 = holes for manipulator handling 4 = Threaded hole (M 8) for remote-controlled connecting the target head with the target support

The described target handling device is operated completely remote-controlled from a control desk situated in the operator's room. The different operations are controlled by optical signals. The system has been in operation for more than two years. It works very satisfactorily and reliably. In our opinion, it is very suitable for routine activation with any small cyclotron.

2.2. Production of ^{67}Ga-citrate and $^{85}SrCl$

The only isotope produced by the Rossendorf cyclotron for sale since 1970 has been ^{85}Sr. As a target a melted RbCl-layer on silverbacking was used for activation with the external 50 μA beam of 14 MeV deuterons.

^{67}Ga has been routine produced since 1978 at our institute. Natural Zn is electroplated on the copper-backing of the target head as shown in Figure 2.2. A mechanical treatment of the Zn-layer after electroplating is necessary. The target thickness is 150—200 μm, irradiations are carried out under a 20° angle at the external deuteron beam directly behind the deflection electrode using the remote-controlled target-handling and irradiating system (Figure 2.1.). Because of the simultaneously synthesized ^{66}Ga, the radiation doses of freshly irradiated targets are extremely high, therefore all irradiations for ^{67}Ga-production are carried out before the week-end shutdown; the radiochemical processing starts on Monday. The irradiated Zn-layer is dissolved in HCl, the ^{67}Ga is extracted from the 7 M HCl solution with isopropylic ether, reextracted with water, evaporated until dryness. The residue is dissolved in an isotonic solution of NaCl-Na-citrate with a pH of 5.5—6.5. After filtration and sterilization the product is ready for i.v. injection.

Table 2.6. shows the development of $^{85}SrCl$ and ^{67}Ga-citrate production volume during the last few years.

2.3. Production of ^{123}I

According to *Sodd* et al., there are three methods for producing ^{123}I meeting the energy conditions of our cyclotron:

^{123}Sb (α, 2n) ^{123}I

^{122}Te (d, n) ^{123}I

^{123}Te (^{3}He, 3n) ^{123}I

With regard to the fairly low intensity of ^{3}He and α-particles at present available, we decided to develop a method for the production of ^{123}I by deuteron reaction. The objective is to obtain activities which are at least sufficient for starting research work in labelling organ-specific radiopharmaceuticals.

We consider the main problem in the ^{122}Te (d, n) ^{123}I reaction to avoid significant losses of the very expansive enriched target material. The choice of target technology and separation procedure meets these conditions.

The target design is drawn in Figure 2.3.

About 220 mg $^{122}TeO_2$ (87 % enrichment) were melted on a platinum backing (1 mm thickness, 22 mm diameter). The target was cooled from the back side by water and at the front side by pressurized air. The separation of

Table 2.6. Summary of routine isotope production at the Rossendorf U-120 cyclotron since 1977.
oIHA – ortho-iodo-hippuric acid;
IHA – 17-iodo-heptadecanoic acid

Year	Product	Shipments	Total activity Bq	mCi
1977	$^{85}Sr\,Cl_2$	77	$3.9 \cdot 10^{9}$	81
	^{67}Ga-cit	—	—	—
1978	$^{85}Sr\,Cl_2$	27	$1.76 \cdot 10^{9}$	47.5
	^{67}Ga-cit	182	$1.89 \cdot 10^{10}$	540
1979	$^{85}Sr\,Cl_2$	56	$3.2 \cdot 10^{9}$	86.5
	^{67}Ga-cit	190	$2.84 \cdot 10^{10}$	760
1980	$^{85}Sr\,Cl_2$	32	$1.24 \cdot 10^{9}$	33.5
	^{67}Ga-cit	168	$3.7 \cdot 10^{10}$	1,000
	^{123}I	—	under investigation	
	^{211}At	—		
	^{81}Rb-^{81m}Kr-generators	—		
1981	$^{85}Sr\,Cl_2$	32	$1.46 \cdot 10^{9}$	395
	^{67}Ga-cit	220	$4.5 \cdot 10^{10}$	1,220
	[^{123}I]-NaI	10	$6.4 \cdot 10^{9}$	170
	[^{123}I]-oIHA	3	$2.4 \cdot 10^{9}$	65
	^{81}Rb-^{81m}Kr-generators	10	$2.0 \cdot 10^{10}$	540
	^{211}At	10	for research only	
1982	$^{85}Sr\,Cl_2$	25	$9.3 \cdot 10^{8}$	25
	^{67}Ga-cit	205	$4.4 \cdot 10^{10}$	1,190
	[^{123}I]-NaI	10	$5.0 \cdot 10^{9}$	135
	[^{123}I]-oIHA	16	$5.9 \cdot 10^{9}$	160
	^{81}Rb-^{81m}Kr-generators	15	$4 \cdot 10^{10}$	1,000
	^{211}At	15	for research only	
1983	$^{85}Sr\,Cl_2$	28	$1.2 \cdot 10^{9}$	31
	^{67}Ga-cit	198	$4.2 \cdot 10^{10}$	1,145
	[^{123}I]-NaI	10	$2.7 \cdot 10^{9}$	72
	[^{123}I]-oIHA	10	$2.5 \cdot 10^{9}$	67
	^{81}Rb-^{81m}Kr-generators	13	$2.7 \cdot 10^{10}$	720
	^{211}At	15	for research only	
	[^{123}I]-IHA		under investigation	
1984	$^{85}Sr\,Cl_2$	26	$9.14 \cdot 10^{8}$	24.7
	^{67}Ga-cit	166	$3.6 \cdot 10^{10}$	985
	[^{123}I]-NaI	14	$4.2 \cdot 10^{9}$	112
	[^{123}I]-oIHA	12	$3.6 \cdot 10^{9}$	96
	[^{123}I]-IHA	11	$2.0 \cdot 10^{9}$	55
	^{81}Rb-^{81m}Kr-generators	21	$3.2 \cdot 10^{10}$	850
	^{211}At	18	for research only	
	[^{18}F]-F_2		under investigation	

the air-stream from the vacuum of the beam line was achieved by a 20 μm steel foil (4); a collimator situated in front of the foils ensured the irradiation of a defined area of the target (11 mm in diameter). The target withstand a 8–10 μA deuteron current of 14 MeV energy for at least 5 hours. A specially designed filter in the air-stream protected the environment from iodine contaminations in accidental cases.

At present the inhomogenity of the beam intensity in the irradiation area is the main problem to be tackled. After solving this problem we will certainly be capable to increase intensity up to 15 $\mu A/cm^2$ or more.

The separation of ^{123}I from the irradiation $^{122}TeO_2$-target has been achieved by thermochromatographic techniques (Figure 2.4.). The target was heated up to 1073 K and after that immediately cooled ensuring a maximal period of 2 minutes above the melting point of tellurium dioxide (1008 K). Under

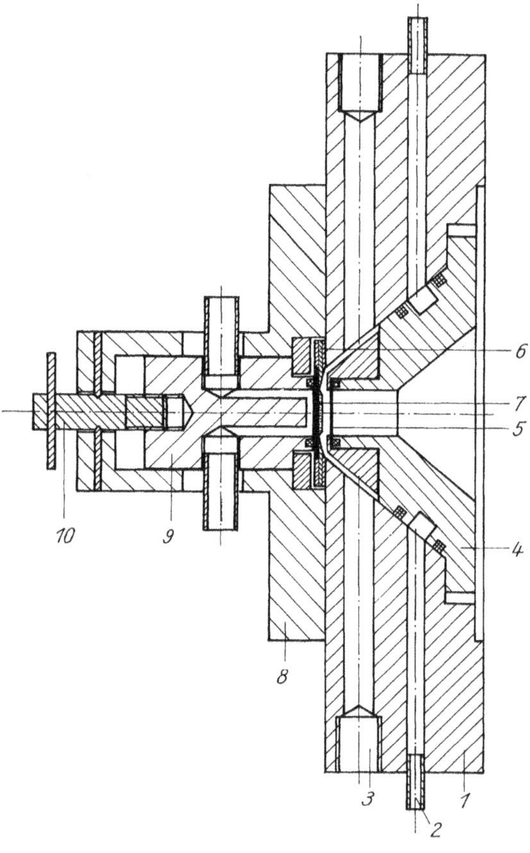

Fig. 2.3. Target irradiating facility for sensitive target materials with low thermo transmission as TeO_2
1 = piacryle flange 2 = cooling water channel 3 = cooling air channel 4 = collimator 5 = 25 μm stainless steel foil 6 = target holder 7 = target coint (Pt with TeO_2) 8 = support 9 = cooling water support 10 = screw

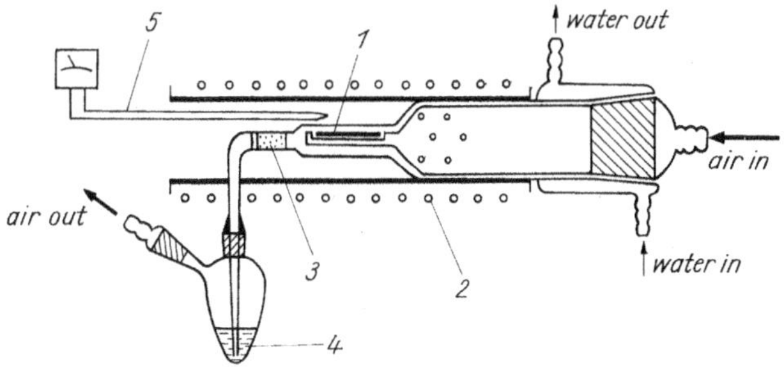

Fig. 2.4. Thermochromatographic equipment (quartz) for ^{123}I-separation from irradiated $^{122}TeO_2$ targets
1 = Pt backing with irradiated $^{122}TeO_2$ 2 = furnace 3 = Al_2O_3-filter 4 = trap
5 = thermocouple

these conditions the loss of valuable material by evaporation could be limited to not more than 1 mg per production cycle. ^{123}I is released from the target with a yield of more than 95 %, and transported by an slow air-stream through an alumina filter to a small trap filled with 0.5 ml of a buffer solution containing $1.9 \cdot 10^{-2}$ M NaHCO + $0.82 \cdot 10^{-2}$ M Na_2CO_3 solution. The alumina filter was used for scavenging traces of tellurium dioxide. In a preliminary study using ^{132}Te-^{132}I as a model system the efficiency of this method was confirmed. Above 873 K all tellurium was retained but 98 % of iodine activity passed the filter. After annealing, the target is ready for the next irradiation without any further treatment. The radioactive purity of the final product was checked by γ-spectroscopy.
Due to the fairly low enrichment of the tellurium-122 an isotopic composition of ^{123}I: 94 %, ^{124}I: 0.7—0.8 %, ^{125}I: 0.1 %, ^{126}I: 0.2—0.3 %, ^{130}I: 4.5 %, ^{131}I: 0.1 % was found at the end of bombardment. ^{123}I-yields between 0.7 and 0.8 mCi/μA (EOB) were obtained. The main impurity was ^{130}I (12 hr). The more dangerous nuclides ^{124}I, ^{126}I were below 1 % at EOB. Further efforts are focussed on overcoming this problem. Despite this unsatisfactory purity, we are using these products for preliminary radiopharmacological studies.

2.4. Production of ^{81}Rb-^{81m}Kr-generators

Yields of approximately 40 MBq/μA h (1 mCi/μA h) were expected for ^{81}Rb production via Br (α, 2n) ^{81}Rb reaction. After completing some improvements resulting in an increase in the α-intensity up to 15 μA, a series of preliminary runs were performed in order to assess the possibilities for routine production of ^{81}Rb-^{81m}Kr generators with our cyclotron. The front side of the target block (silver) was furnished with grooves (0.3 mm deep) to improve the adhesion of a salt layer. NaBr was melted on to this surface.
Silver was used to avoid difficulties by surface oxidation during the plating procedure.

Table 2.7. Typical composition of the target-solution for a five-hour NaBr-activation

	$T_{1/2}$	(Bq/μA)	(μCi/μA)	(%)
^{81}Rb	4.58 h	$3.1 \cdot 10^7$	840	82.7
^{82m}Rb	6.4 h	$6.3 \cdot 10^6$	170	16.7
^{83}Rb	83 d	$1.5 \cdot 10^5$	3.9	39
^{84}Rb	34.5 d	$5 \cdot 10^4$	1.2	12
^{22}Na	2.6 a	$3.7 \cdot 10^3$	1	1
^{24}Na	15 h	$5 \cdot 10^4$	13	13

Irradiations have been carried out under surface/beam angles of 20 degr, and 15 μA 28 MeV α-particles for 1—5 hours.
The irradiated NaBr was dissolved in 25 ml water. The isotopic composition was determined γ-spectroscopically.
The typical composition of the target-solution for a five-hour NaBr-activation was the following (at EOB; Table 2.7.).
According to these figures, more than 0.4 GBq (100 mCi) ^{81}Rb can be produced with α-particle currents of 15—20 μA. Thus, the production of at least two generators per run should be possible.
The NaBr-solution was filtered through a G-5 glass sinter funnel (Figure 2.5.) and poured into a column (30 $\times$ 5 mm) filled with the inorganic cation-exchanger BIO-RAD ZP-1 (20—50 mesh). The elution rate was 10 ml/min. Under these conditions more than 95 % of the ^{81}Rb-activity were retained by the column. After washing with 100 ml of distilled water, the sodium ions were almost completely displaced.
^{81m}Kr-gas was extracted by flushing the column with a continuous air-stream of 3 l/min. About 50 % of the equilibrium in dextrose are planned.

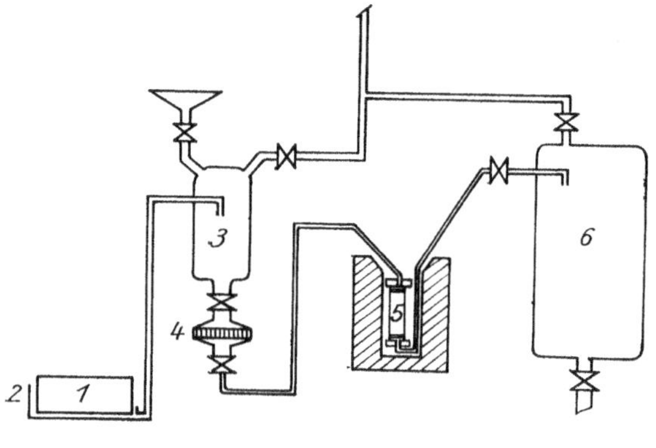

Fig. 2.5. Scheme for the production technology of ^{81}Rb-^{81m}Kr-generators
1 = target head 2 = vessel for dissolving irradiated NaBr 3 = storage vessel 4 = filter 5 = generator column (Zr-phosphate) 6 = waste vessel

2.5. Production of ^{211}At

Recently we started investigations into astatine chemistry using the irradiation possibilities at the Joint Institute of Nuclear Research Dubna URS. After the 680 MeV synchrocyclotron ceased its operation, which was due to reconstruction-work, the question arose to provide another source for this radioisotope.

^{211}At is rather selectively produced at reasonable yields (0.4 mCi/μA h). Contaminations of the by-product ^{210}At being generated by the $^{209}Bi(\alpha, 3n)^{210}At$ reaction strongly depend on α-energy. Under these conditions (28 MeV α-particles) only very small ^{210}At contaminations were predicted. A bismuth layer of 20 μm thickness and 20 cm^2 effective area deposited by means of vacuum-evaporation was irradiated; the effective target thickness increased to 110 μm. The target was situated directly behind the deflection electrode. After 1–3 hours of irradiation at 3 μA no serious astatine release and surface erosion could be detected.

The integral dose rate of the irradiated target was 15 rem/h in a distance of 10 cm. The main reason for this high radiation dose was the activation of the copper behind the bismuth layer. Therefore, the bismuth was scrapped mechanically from the target holder and processed by thermochromatography.

The bismuth was oxidized by air at 950 K. The released At was quantitatively absorbed on a silver surface maintained at a temperature of (450 $\pm$ 20) K.

Impurities such as ^{66}Ga, ^{67}Ga, ^{111}I are retained completely in the resulting Bi_2O_3.

An alternative route was the absorption of the volatilized astatine directly in 0.2 ml water.

Thus, ^{211}At-amounts of more than 10 mCi are routineously available opening the possibility of inorganic, organic chemical and nuclear medical research work.

2.6. Other possibilities for isotope production

A number of modern radioisotopes principally available at the Rossendorf U-120 cyclotron are listed in Table 2.8. Under these conditions the main nuclear reaction to produce such isotopes will be the (α, 2n) reaction.

Despite a sufficient increase in α-beam intensity it would be impossible to meet the whole demand of the GDR. The only exceptions will be ^{67}Ga, ^{81}Rb and perhaps ^{111}In. Our cyclotron provides ideal conditions to produce the ultra-short β^+-emitting isotopes including ^{18}F for synthesizing ^{11}C- or ^{18}F-labelled metabolic or pharmaceutic compounds. The short-lived β^+-emitting isotopes may be produced not only by cyclotrons, in some cases even by electrostatic accelerators.

Table 2.8. Choice of modern radioisotopes principally available at the Rossendorf U-120 cyclotron

Nuclide	Half-life	Radiation	Production reaction possible with	Remarks
^{11}C	20 m	β^+	^{11}B (p, n)	off-line only
			^{10}B (d, n)	^{11}CO
			^{11}B (d, n)	$^{11}CO_2$
			^{14}N (p, α)	on-line 5.10^{10} Bq
^{13}N	10 m	β^+	CH_4 (d, n)	on-line 5.10^{9} Bq
^{15}O	123 s	β^+	^{14}N (d, n)	on-line
^{18}F	110 m	β^+	^{16}O (α, pn)	water-target
			^{20}Ne (d, α)	on-line, off-line 10^{11} Bq
^{52}Fe	8.3 h	β^+ (58 %)	^{50}Cr (α, 2n)	4.10^{9} Bq
		169 keV 100 %		
^{67}Ga	78 h	93 keV 38 %	Zn (d, xn)	4.10^{9} Bq
		185 keV 20 %		
		300 keV 15 %		
		394 keV 4 %		
^{77}Br	56 h	249 keV 3 %	^{76}Se (d, n)	
		297 keV 5 %	^{75}As (α, 2n)	5.10^{8} Bq
		521 keV 24 %		
^{81}Rb-	4.58 h		^{79}Br (α, 2n)	5.10^{9} Bq
^{81m}Kr	13 s	190 keV 67 %		
^{85}Sr	64 d	514 keV 99 %	^{85}Rb (d, 2n)	10^{8} Bq
^{87}Y-	80 h	485 keV 96 %	^{85}Rb (α, 2n)	5.10^{8} Bq
^{87m}Sr	2.8 h	388 keV 83 %		
^{111}In	2.8 d	171 91 %	^{110}Cd (d, n)	4.10^{9} Bq
		245 94 %	^{109}Ag (α, 2n)	10^{9} Bq
^{123}I	13.2 h	159	^{122}Tc (d, n)	2.10^{9} Bq
			^{121}Sb (α, 2n)	2.10^{9} Bq
			^{123}Te (p, n)	tandem
^{129}Cs	32 h	372 32 %	^{127}I (α, 2n)	4.10^{8} Bq
		411 22 %		
^{157}Dy	8.1 h	326 95 %	^{155}Gd (α, 2n)	10^{8} Bq
^{211}At	7.2 h	α, K_X-Po	^{209}Be (α, 2n)	5.10^{8} Bq

References

1. *Clark, J. C., P. D. Buckingham:* Short-lived radioactive gases for clinical use. Butterworth Publishing Company Ltd, London 1975.
2. *Nättö, V., E. M. Suolinna* (eds.): Medical application of cyclotrons II. Proceedings of the Third Symposium on Medical Application of Cyclotrons, June 13–16, 1983, Turku, Finland.
3. *Quaim, S. M., G. Stöcklin:* Production of some medically important short-lived neutron deficient radioisotopes of halogens. Radiochim. Acta **34** (1983), 25.
4. *Lambrecht, R. M.:* Positron emitting radionuclides—present and future status. Radiopharmaceuticals II, Society of Nuclear Medicine New York (1971), 753.

5. *Ruiz, Vera:* Cyclotron facilities with radionuclide production programmes in member states, a compilation of the IAEA. IAEA, Vienna, January 1983.
6. *Vaalburg, W., A. M. J. Paans:* Short-lived positron emitting radionuclides. Radionuclides Production II (F. *Helus*, ed.), CRC Press, Florida (1983), 47.
7. *Waters, S. L., D. J. Silvester:* Inorganic cyclotron radionuclides. Radiochim. Acta **30** (1982), 163.
8. *Welch, M. J.* (ed.): Radiopharmaceuticals and other compounds labelled with short-lived nuclides. Special issue of Int. J. Appl. Radiat. Isotopes **28** (1977), 1.
9. *Wolf, A. P., W. B. Jones:* Cyclotrones for biomedical radioisotope production. Radiochim. Acta **34** (1983), 1.

Chapter 3

Radionuclide Generator Systems

Harald Deckart

The "generator" (called "cow") contains a parent radionuclide, which is reactor-activated, with a physical half-life long enough to be sent to the users and to generate there a "daughter" radionuclide with a short half-life. This can be eluted from the parent radionuclide contained in the column every day just before it is used in the nuclear medicine department ("milking the Cow").

Figure 3.1. shows the principle of a generator system: the adsorber column (fixed in lead-shielded glass vial) contains the parent radionuclide (see Table 3.1.). Its upper part is connected with a vial containing a sterile elution liquid (physiologic saline, acidic solution); its lower part is connected with a vial which takes up the eluate, containing the daughter radionuclide; this is nowadays evacuated and can quickly withdraw the eluate by its pressure difference. The elution process can be repeated several times depending on the half-life of the parent nuclide. Winsche et al. /156/ described the first ^{132}I-generator system in 1951. The worldwide used $^{99}Mo/^{99m}Tc$-generator was developed in 1957 /123/ and introduced into clinical practice by Harper in 1962 /62, 63/, the $^{113}Sn/^{113m}In$-generator by Subramanian in 1967 /128/.

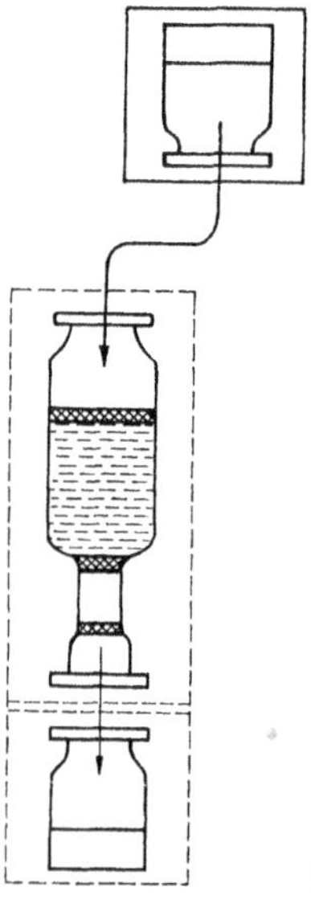

Fig. 3.1. The principle of a generator system

Table 3.1. Clinically used generator system – Clinical application.
1 – brain scintigraphy; 2 – bone scintigraphy; 3 – lung perfusion; 4 – lung ventilation; 5 – myocardial perfusion; 6 – radiocardiography, ventriculography; 7 – blood volume determination; 8 – kidney perfusion; 9 – thyroid scintigraphy; 10 – thyroid uptake; 11 – tumor scintigraphy; 12 – treatment; 13 – see text; 14 – nowadays rarely used or abandoned

Parent			Daughter						References
Nuclear reaction	Symbol	half-life	Symbol	Half-life	Principal photon energy (MeV)	(% abundance)	Decay into stable nuclide	Clinical application	
a) Fission	^{99}Mo	67 h	^{99m}Tc	6.0 h	0.140	(90)	^{99}Tc (Tp 2.1 · 10^5y β^-: 0.292) —— ^{99}Ru		/ 62/
b) ^{98}Mo (n, γ)									
^{112}Sn (n, γ)	^{113}Sn	115 d	^{113m}In	1.7 h	0.393	(64)	^{113}In	13	/126/
Fission	^{132}Te	3.2 d	^{132}I	2.4 h	0.24–1.4 β^-: 2.1	(14)	^{132}Xe	10, 14	/154/
^{114}Cd (n, γ)	^{115}Cd	2.3 d	^{115m}In	4.5 h	0.335 β^-: 0.840	(50)	^{115}In	14	
Fission	^{137}Cs	30 y	^{137m}Ba	2.6 min	0.662	(85)	^{137}Ba	1, 6 7, 11	/ 17/
^{197}Au (p, 3n)	^{195m}Hg	41.6 h	^{195m}Au	30.5 s	0.068 0.261	(77)	^{195}Au	6	
a) ^{85}Rb (p) b) ^{85}Rb (p, 5n) c) ^{79}Br (4 He, 2n) d) ^{80}Kr (4 He, n)	^{81}Rb	4.6 h	^{81m}Kr	13 s	0.191	(65)	^{81}Kr	3, 5	/156/
^{69}Ga (2p, 2n)	^{68}Ge	280 d	^{68}Ga	68 min	0.511 (β^+)	(176)	^{68}Zn	1, 2, 7, 11	/ 88/
	^{87}Y	80.3 h	^{87m}Sr	2.81 h	0.388	(80)	^{87}Sr	2, 14	/ 2/
	^{90}Sr	28.5 a	^{90}Y	64.1 h	no γ β^-: 2.27 MeV		^{90}Zr	12	/ 39/

The dynamic process of activity in the daughter nuclide is characterized mathematically in Chapter 1 (Appendix: after elution the activity of the daughter radionuclide again is slowly growing until an activity equilibrium between parent and daughter is achieved: elution of activity can be made repeatedly; it can be made even before the equilibrium is achieved with a lower yield. Without elution both decay with the physical half-life of the parent nuclide (Figure 3.2.).

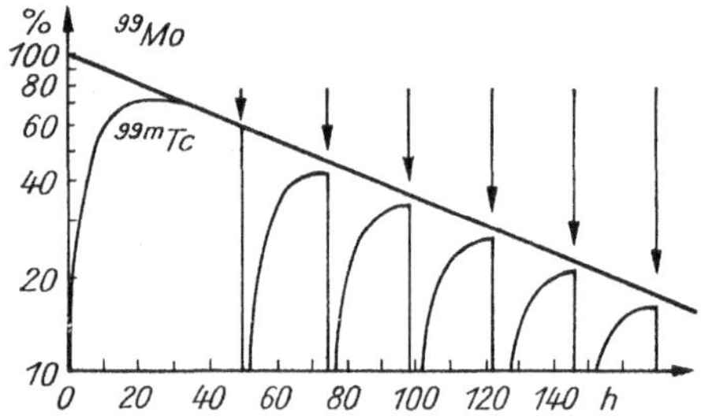

Fig. 3.2. Decay of the parent nuclide and elution of daughter nuclide

1. The technetium-generator

The parent nuclide can be produced by

– ^{98}Mo *(n,γ)* ^{99}Mo *reaction.* A molybdenum target irradiated with neutrons contains naturally occurring ^{98}Mo, or is composed of different organic molybdenum complexes. After neutron-irradiation in the reactor an enrichment of ^{99}Mo occurs, thereafter molybdate is separated in different ways. The carrier containing reactions leads only to low specific activities.

– ^{235}U (n,f) ^{99}Mo
^{238}U (n,f) ^{99}Mo. This *fission-produced molybdenum* is separated by acidic solutions and passed through exchange resin columns. The fission-produced molybdate is carrier-free and yields high specific activity. Problematic is the separation of radionuclide impurities.

2. Types of generator systems

Solvent-extraction generator:
^{99}Mo-alkaline solution is separated from ^{99m}Tc-pertechnetate by liquid-extraction with methylethylketone (MEK), thereafter the MEK is distilled. With this technique one can obtain high activity concentrations of pertechnetate free from contaminants from relatively low activity molybdate; it is time-consuming to operate this generator.

Sublimation generator:
A stream of oxygen is passing the heated and activated molybdenum trioxide. Technetium heptaoxide is evaporated and condensed in a cooled region of the system, from there it is eluted by saline or extracted by MEK. Advantage and disadvantage are the same as in the solvent-extraction generator.

Chromatographic generator:
This system is the most widely used device. Aluminium oxide (Al_2O_3) is used as sorbent of parent molybdate $^{99}MoO_4^{2-}$).

Column-adsorber material is aluminium oxide for the Mo/Tc-generator, physiologic saline is used as eluant; the formerly used diluted acidic solutions led to elution of sorbent-material. Hydrous zirkonium oxide (ZrO_2) or organic exchange resins are used in the Sn/In-generator; elution with weak hydrogen chloride solution. In the $^{195m}Hg/^{195m}Au$-generator ZnS is the sorbent, sodium thiosulphate solution is used as eluant; repeated elutions are every 3—5 minutes possible.

3. Quality control in the generator system

Elution profile: If one elutes the generator, the activity increases at the beginning, reaches a maximum, and decreases thereafter (Figure 3.3.). To obtain highly concentrated activity for labelling a fractionated elution is

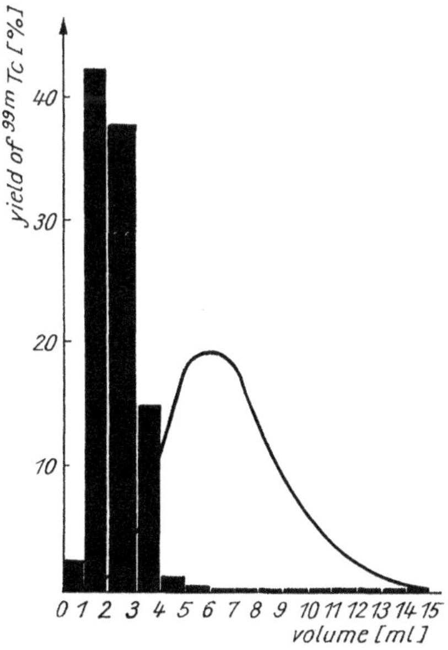

Fig. 3.3. Elution profile
columns — fractionated elution
curve — continuous elution

recommended (Figure 3.3. too). Using a carrier-free fission-produced ^{99}Mo-generator one obtains smaller volumes with high activity in contrast to the mostly purchased ^{99}Mo (n,γ)-reaction generator.
Every generator should be washed thoroughly once or twice before use to remove impurities. The elution yield amounts to about 75—85 %; it varies from generator to generator and can be influenced by defects in the adsorber column.

Chemical purity: Sorbent material is retained by a fritte. Its contamination of the eluate occurs after exsiccation of the column: Al is found together with Al_2O_3, which is building colloids at pH 5 and can interfere in the Tc-labelling procedure. It may cause agglutination in Tc-S-colloid or in Tc-RBC. With a colour spot test with paper containing quinalizarin compared to a standard aluminium solution contents > 0.005 mg may be detected. Breakthrough of zirconia-sorbent in ^{113}Sn-generator is detected qualitatively with 0.05 %-aqueous alizarine solution: one drop of eluate (< 0.7 μg zircon/ml) leads to a yellowish colour. Quantitative photometric measurements are performed with Morin-ethanol-solution.
Radiochemical purity: the Tc-eluate should contain 99 % TcO_4^- (VII). Differentiation of the lower valence Tc is performed with paper chromatography (PCP), Sephadex-G 25—column chromatography, thin-layer chromatography (see Chapter 17).

Radionuclide purity: depends on the purity of the target material, the type of its irradiation and the quality of separation techniques (destillation, ion exchange). Optimal control is ensured by Ge-Li-spectrometry with multichannel pulse-height-analyzer. The tin-113-generator has ^{113}Sn, ^{117m}Sn, ^{125}Sb, as contaminants; in fission-produced ^{99}Mo-generators ^{134}Cs, ^{131}I, ^{132}Ru, ^{124}Sb, ^{103}Ru, ^{99}Zr, ^{86}Rb may be contaminants in the eluate, ^{186}Re, ^{186}Re in neutron-irradiated molybdenum.

^{99}Mo-breakthrough should be < 2 μg/ml eluate. A qualitative colour test is carried out with strips soaked in $ZnSO_4$-solution, after drying it is impregnated with concentrated sodium-xanthat-solution; reddish colouring occurs if the eluate contains > 0.01 μg ^{99}Mo.

^{113}Sn-breakthrough: qualitative detection of tin atoms: there is a blue colouring if 30 μg Sn/ml eluate is put into 0.1 % alcoholic hematoxyline solution.

References: see Chapter 5.

Chapter 4

Radiopharmaceuticals: A Review

Harald Deckart

This chapter deals with radiopharmacy, radiopharmacology and the clinical application in diagnostics and treatment in the past, currently and with the development of some new compounds with a prospective use in the near future. The following radionuclides and compounds labelled with them will be discussed:

^{132}I,	^{131}I	^{201}Tl	^{133}Xe		^{64}Cu
^{125}I,	^{123}I	^{203}Hg, ^{197}Hg	^{90}Y		^{58}Co, ^{57}Co
^{113m}In,	^{111}In	^{198}Au	^{90}Sr,	^{89}Sr, ^{87m}Sr, ^{85}Sr	^{59}Fe
^{99m}Tc		^{185}Re	^{81m}Kr		^{51}Cr
		^{169}Er	^{75}Se		^{32}P
		^{169}Yb	^{68}Ga,	^{67}Ga	^{18}F

^{18}F, ^{57}Co, ^{67}Ga, ^{85}Sr, ^{111}In, ^{123}I and ^{201}Tl are cyclotron produced radionuclides (see Chapter 2). ^{68}Ga, ^{81m}Kr, ^{87m}Sr, ^{90}Y, ^{99m}Tc, ^{113m}In and ^{132}I can be eluted from parent/daughter radionuclide containing generator-systems (see Chapter 3): all the other radionuclides including parent nuclides are reactor produced (see Chapter 1).

1. Methods of labelling

Isotope exchange

Inactive atoms in the molecule are replaced by the radioactive ones of the same isotope, especially used in iodination (see *I-Rose bengal, *I-hippurate, *I-diatrizoate) and in labelling with tritium ^{3}H:

$$R^{127}I + {}^{131}I^{-} \rightleftharpoons R\text{-}^{131}I + {}^{127}I^{-}$$

The velocity of exchange depends on temperature, concentration, solvent and pH of the medium. With a nearly carrier-free radio-iodide a high labelling yield of the same microgram-weighing substance can be obtained. The labelled substance has the same biochemical properties as the unlabelled.

Iodination

For iodination iodide (I^{-}) must be transferred to its oxidized form (Iodine: I_2). The following two methods are commonly used:

Iodine monochloride method /94/

ICl is used instead of I_2. $^{+}I^{-}$ is exchanged with ^{127}Cl in ^{+}ICl in diluted HCl. Another possibility is the production of ^{+}ICl by:

$$2\ Na^{+}I + NaIO_3 + 6\ HCl \rightarrow 3^{+}ICl + 3\ NaCl_3 + 3\ H_2O$$

Then this solution is put together with the molecule to be labelled by isotope exchange; yields between 50–90 % are obtained. This method can be used also as labelling technique for aromatic compounds and for addition reactions of double bonds, e.g. the labelling of fatty acids and oily compounds.

Chloramine T-method /56/

Chloramine T oxidizes iodide to iodine which labels the molecule (I^+-substitution tyrosyl-groups) with high yields of about 100 %. This previous method has been actually modified and is used nowadays with minimal excess of chloramine-T and a reaction time of only some seconds. In this modification it is as gentle as electrolytic method and enzymatic method for the labelling of proteins without damaging the molecule.

Electrolytic method /113/

K*I-solution and the labelling material are brought into the anode compartment of an electrolytic system. During electrolysis iodine is continuously deliberated for iodination of the compound (the tyrosyl-groups of protein), which has no contact with oxidizing or reducing material with yields of about 80 % in only moderately specific activities.

Enzymatic method /101, 142/

Enzymes (lactoperoxidase, glucoseoxidase, chlorperoxidase) are added to the mixture of *iodine and the agent to be labelled. Iodide is oxidized by H_2O_2 to iodine which reacts with the protein. This method is a mild procedure with yields of about 60–80 % and higher specific activity. In proteins particularly the tyrosylgroups were iodinated stable, or the imidazole ring of histidine. Reversible binding occurs to amino- or sulphhydryl groups. The binding to aromatic compounds is irreversible, to aliphatic chains in vivo more unstable.

Foreign labelling

If the pharmaceutical has no element (especially I) in its molecule which is exchangeable with a radioactive isotope of this element, this element or a group containing the label has to be incorporated. Labelling of antigens (amine acids, proteins, peptides) occurs in a substitution in the aromatic ring of tyrosine.

Biosynthesis /14/

Some compounds with a complicated structure can be labelled during their biosynthesis: Vitamin B_{12} is synthesized by bacterial cultures of Streptomyces griseus with addition of ionogen ^{60}Co, ^{58}Co or ^{57}Co. ^{75}Se-selenomethionine is synthesized by cultures of Saccharomyces cerevisiae with addition of ^{75} Se-sodium selenite which replaces sulphur atoms without changing its biological behaviour.

Procedures for stabilizing radiopharmaceuticals

After iodination the reaction mixture containing free iodide must be separated by precipitation, dialysis, gel filtration, and anion exchange. Sterilization is performed by autoclaving, in protein only by filtration through bacterial filters.

Industrially prepared and commercially distributed, radiopharmaceuticals often contain bacteriostatic agents as benzyl alcohol (0.9 % concentration). For stabilizing reasons pH of the solution is very important, therefore acid or alkali buffers are added, mostly tris- or phosphate buffers are used. To avoid oxidation of iodide reducing agents (sodium thiosulphate, ascorbic acid, sodium sulphite) are often added. To stabilize colloid preparations "stabilizers" are added: gelatine, polyvinylpyrolidone, Tween 80 (citrate, ascorbate, ascorbic acid). Some of these stabilizers seem to be responsible for adverse reactions, e.g. gelatine, Infukoll®, Haemacel® (see Chapter 18).

2. Radioiodine. *I-labelled compounds

About 5 out of 24-iodine-isotopes have been used in nuclear medicine (Table 4.1.): ^{131}I, ^{125}I and ^{123}I are the most important today. Iodine is a nonmetallic element belonging to the halogen-group VIIb in the Periodic Table with

Table 4.1. Radionuclides in different chemical forms used as radiopharmaceuticals

Inorganic cations	^{24}Na-, ^{42}K-, ^{45}Ca-, ^{47}Ca-chloride ^{59}Fe-citrate, ^{64}Cu-, ^{81}Rb-, ^{86}Rb-chloride ^{85}Sr-, ^{87m}Sr-, ^{89}Sr-, ^{90}Sr-, ^{131}Cs-chloride, ^{197}Hg-, ^{201}Hg-, ^{201}Tl-chloride
Inorganic anions	^{18}F-fluoride, ^{77}Br-bromide, ^{32}P-phosphate, ^{35}S-sulphate, ^{51}Cr-chromate, ^{123}I-iodide ^{125}I-, ^{131}I-, ^{132}I-iodide
Organic molecules	^{131}I-, ^{113m}In-, ^{99m}Tc-HSA, ^{131}I-fibrinogen, ^{111}In-fibrinogen, ^{99m}Tc-plasmin
Organic molecules substitution labelled	^{75}Se-selenomethionine, ^{57}Co (^{58}Co-)bleomycine, -cyanocobalamine, ^{131}I-BSP, -Rose bengal, -hippurate
Aggregation and deformation of organic molecules (particles)	^{131}I-, ^{123}I-, ^{111}In-, ^{99m}Tc-labelled microaggregates (diameter 10—35 μm) macroaggregates (diameter 35—65 μm) microspheres (diameter 1—10, 10—35, 35—65 μm)
Colloids	^{198}Au-colloid, ^{90}Y-citrate colloid, ^{169}Er-citrate colloid, ^{186}Re-sulphide colloid, ^{99m}Tc-sulphide colloid
Labelled blood cells	^{51}Cr-, ^{111}In-, ^{113m}In-, ^{99m}Tc-labelled red blood cells (RBCs), labelled, heat damaged RBC ^{51}Cr-, ^{111}In-leucocytes, -lymphocytes, -platelets
Complexes	all ^{99m}Tc-compounds, all ^{111}In-, ^{113m}-compounds, ^{67}Ga-citrate, ^{169}Yb-DTPA
Aerosols	^{99m}Tc-DTPA, ^{99m}Tc-S-colloid nebulized (particle size: 1—3 μm)
Gases	^{133}Xe, ^{85}Kr, ^{81m}Kr, ($^{11}CO_2$, ^{11}CO, $^{13}N_2$, $^{15}O_2$, $C^{15}O_2$, C^{15}O)
Water	^{3}H (tritiated)-water

fluorine, bromine, chlorine and astatine. Iodine is used in its oxidation state − 1 as iodide (I^-) for thyroid diagnosis and treatment, for labelling organic compounds in the oxidation state + 1, − 1 /68/.

^{123}I has ideal physical properties for scintigraphic resolution with a monoenergetic γ-line (0.159 MeV (84 %₀ abundance)); no β-component together with the short half-life (Tp 13.3 h) leads to a low radiation exposure per mCi; repeated examinations are possible. Depending on its cyclotron production ^{124}I-, ^{126}I-impurities may be present /98/.

^{125}I is reactor-produced. Tp 60 d, low γ-energy (0.035 MeV, one tenth of ^{131}I γ-line!; X-ray 0.027 MeV) leads to high tissue absorption with a β-equivalent relatively high radiation dose and a possible use for treatment. For in vivo examination it can be used only for superficial organs and tissues; it is mostly used as a label for in vitro tests: long life and low autoradiolysis make it an economical tracer /97/.

^{131}I is reactor produced. Tp: 8.04 d; γ-energy: 0.364 MeV (80 %), 0.637 MeV (9.0 %), further γ-lines with low (< 9 %) abundance; β: 0.608 MeV (87.2 %), further energies (0.8; 0.3; 0.2 MeV) with low abundance. The β-radiation make it usable for treatment purposes and leads to a relatively high radiation dose used as a tracer for diagnostic purposes: It belongs to the radiation toxicity group 2 because of its transition into the gaseous phase and the possible inhalation and thyroid retention by patients and laboratory staff.

^{131}I is available carrier-free and is supplied as sodium iodide in water-clear solution for oral use, in pyrogen-free, sterile solution for intravenous application. For oral use it is available in gelatine-capsules (fixed on the inner wall); sometimes enteral resorption may be enhanced in existing solubilizing problems of the capsule material. *Application* to radioiodine uptake test and (in the past) to thyroid scanning, treatment of benign thyroid diseases (hyperthyroidism, adenoma, euthyroid goiter, recurrent goiter after surgical resection), radiation resection of thyroid remnants after total thyroidectomy in differentiated thyroid carcinoma and of its metastases /68, 69/.

^{132}I (see Chapter 3).

^{131}I (^{125}I)-albumine

Iodination of human serum albumine (HSA; abbrev. RIHSA radioiodinated HSA) must be performed carefully by different methods without damaging the protein molecule to preserve its physiological and biological behaviour (chloramine-T-method, electrolytic technique). Usually one I-atom is bound to one albumin molecule: monoiodotyrosine derivative. Exceeding iodide is removed by sterile anion-exchange resin filter, sterilization by passing through millipore filter, dilution with unlabelled albumin solution. The product (spec. activity 1—10 mCi/mg HSA) is a clear or opalescent solution. Storage (+ 2 to + 4 °C) until 4 weeks. In quantitative measurements only fresh-prepared products should be used, or anion-exchange of deliberated unbound iodide autoradiolysis must be performed /7, 59/.

Application to blood volume determination, cardiac output, peripheral circulation studies and, in the past, to blood scintigraphy and cisternography.

^{131}I-HSA-particles /13/

After labelling albumin with ^{131}I denaturation is achieved by heat (10 min shaking in a water-bath). Particle size depends on pH: adjusting to pH 7.5 leads to *microaggregates* (particle size $< 5\ \mu m$), pH 5.5 to *macroaggregates* (particle size 10—19 μm, abbrev. MAA). These aggregates are non-uniform in shape. Nowadays *microspheres* (human serum albumin microspheres, HAM) with a uniform spherical shape and a rigid and dense structure are usually preferred /11, 27, 71, 78, 103, 134, 160/.

a) Formation of albumin droplets (stirring of human albumin solution in heated cotton-seed oil at + 135 to + 180 °C for 10 min and cooled to + 18 °C).
b) Dilution and washing with ethyl ether or heptan to remove oil.
c) Isolation of microspheres by filtration or mechanical separation by centrifuge according to different particle sizes and drying.
d) Labelling with ^{131}I, ^{111}In, ^{99m}Tc /13—18, 49/.

Application to scintigraphy of RHS of liver and spleen (particle size 10 to 30 μm), bone marrow (particle size 1—10 μm); for lung perfusion scanning (particle size 30—60 μm) or regional blood flow determination (e.g. cardiac muscle in connection with X-ray coronarography), to the detection of arteriovenous shunts in peripheral vascular disease, in liver cirrhosis in combination with splenoportography, to the observation of esophago-gastrointestinal kinetics and determination of gastric emptying rate. MAA and HAM are fixed in thrombotic processes (phleboscintigraphy) and used for localization of deep thrombosis and collateral circulation. For microembolization only 0.3 to 0.5 mg HSA containing 300,000 to 500,000 particles were introduced, which lead to passager embolization of only 0.5 % of the capillary network. Metabolic degradation, phagocytosis and renal elimination of denaturated protein of MAA is quicker than those of microspheres. During metabolization of greater (lung-)particles liver may be imaged transiently with those degraded smaller particles. Using labelling kits one always has to take the same volume and activity to receive comparable results in different patients.

^{131}I (^{125}I, ^{123}I)-Fibrinogen

for thrombosis diagnosis in follow-up measurements (fibrinogen uptake test) or scintigraphy in superficial (^{115}I) or deep (^{131}I) vein thrombosis /94/.

^{131}I (^{123}I)-Streptokinase /50/

This enzyme activates the thrombus-clotting process.
Application to detection of thrombus in statu nascendi /120/.

^{131}I (^{125}I, ^{123}I-sodium orthoiodo-hippurate (OIH) (Figure 4.1.)

Labelling by the isotope exchange method: heating (+ 100 °C) of o-iodohippurate solution together with *I-sodium at pH 4—6; anion-exchange of unbound tracer. Colourless clear solution. ^{131}I leads to autoradiolysis during storage to a higher degree than ^{125}I; in quantitative measurements unbound iodide has to be removed before use (quality control!). Hippurate is excreted by tubular secretion with an extraction efficiency of 65—80 %; no tubular reabsorption /135, 144, 157/.

^{131}I-m-iodo-hippurate has been found to be more stable in vivo compared to the ortho-compound /45/. *Application:* renography (^{131}I for adults, ^{125}I for children; ^{123}I-hippurate reduces the radiation dose to about 3 % compared to ^{131}I-label). Sequential kidney scintigraphy, clearance determinations for quantification of tubular secretion; cystoureteral, pelvic reflux studies.

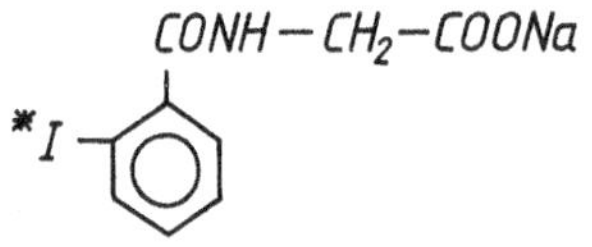

Fig. 4.1. ^{131}I-sodium-orthoidohippurate

^{131}I-Diatrizoate (Figure 4.2.)

The X-ray contrast medium 3,5-dacetylamino-2,4,6-triiodobenzoic acid as sodium salt (Hypaque®) or methylglucamine (Renografin®, Visotrast®) is excreted through kidneys by glomerular filtration. Iodination with iodine exchange method /18, 135/.

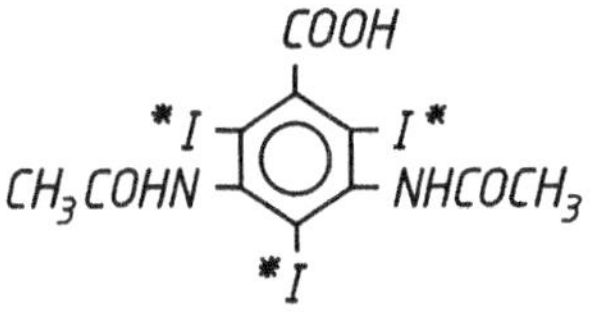

Fig. 4.2. ^{131}I-Diatrizoate

^{131}I-(^{123}I)-Rose bengal (RB) /25, 26/
and ^{131}I-(^{123}I)-Bromsulphthalein (BSP) /27, 28/ (Figure 4.3)

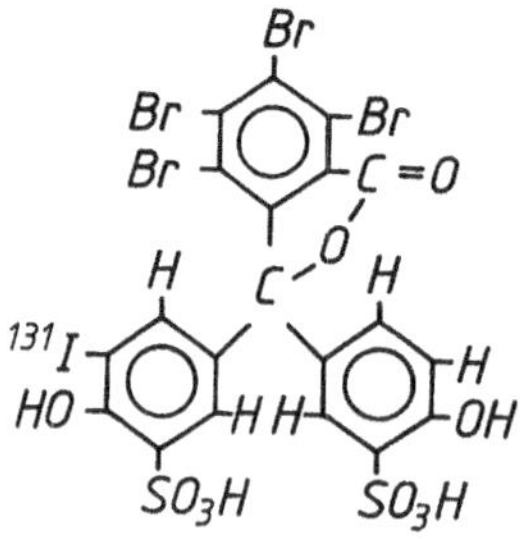

Fig. 4.3. ^{131}I-Bromsulphthalein (BSP)

Rose bengal (= 2,4,5,7-tetraiode, 3',4',5',6'-tetrachlorofluorescein in disodium salt solution) can be labelled by the isotope exchange method or with potassium iodate as oxidant and ^{131}I-sodium iodide; clear deep-red solution which is sensitive to light and autoradiolysis /117, 133/. Because of the variability of the degree of halogen-content, it was replaced in the last decade by BSP (Figure 4.3.), which results in better reproducible iodination. RB is excreted unchanged through liver cells via bile, whereas BSP is coupled by cystein and glutathion /54, 145/. *Application* functional hepatic scintigraphy formerly, at present for to liver function determination by blood clearance techniques or external measurements.

^{131}I-Iodocholesterol (Figure 4.4.)

is commercially available deep-frozen in dry ice. ^{131}I-19-iodocholesterol which was labelled at first was in vivo unstable /29/. Commercially available is 6β-iodomethyl-19-norcholest-5(10-en-3β-ol labelled with ^{131}I or 6-β-methylselenomethyl-19-norcholest-5(10)-en 3β-ol labelled with ^{75}Se; it is accumulated in the adrenal cortex as a precursor of cholesterol synthesis /24, 114/. Adverse reactions immediately after injection have been observed (see Chapter 18). *Application:* Adrenal scanning 3 to 5 days after injection for differential diagnosis of cortex hyperplasia, cortex adenoma (Conn's syndrome), differentiated adrenal carcinoma, recurrence of adrenal tissue after adrenalectomy in Cushing's syndrome.

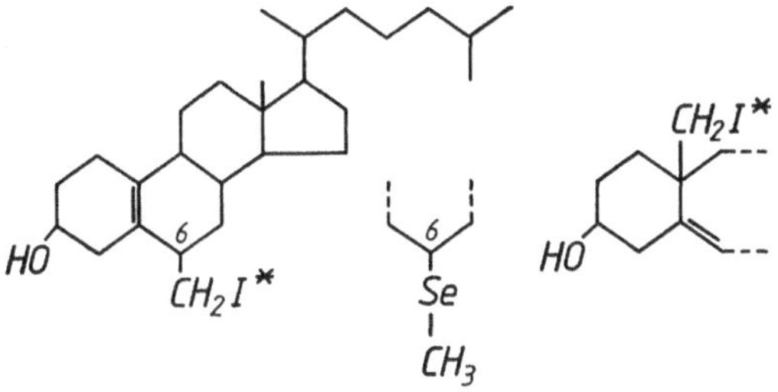

Fig. 4.4. Derivatives of cholesterol

^{131}I-(^{123}I)-meta-iodobenzylguanidine (MIBG)

MIBG is specifically trapped by adrenal medulla celis, as it resembles norepinephrine in molecular structure. It probably enters adrenergic tissue by the same mechanism as that of a neurotransmitter; it is concentrated in the adrenal catecholamine-storage vesicles. Threecyclic antidepressive drugs reduce accumulation of MIBG. Scintigraphy has to be performed repeatedly 1 up to 3 days after injection /152/.

Application: Scintigraphy of adrenal medullar hyperplasia, benign adrenal and extraadrenal primary and malignant metastatic pheochromocytomas, scintigraphy of neuroblastomas and paraganglioma. Trials for treatment have been performed.

^{131}I-Iodoquinoline

Different iodoquinoline derivatives exist labelled with ^{131}I, ^{125}I or ^{123}I and were used for detecting melanoma in eye, skin and melanoma metastases. Because of highly specific uptake it can be used even for treatment /10/.

^{131}I-(^{125}I)-labelled monoclonal (polyclonal) antibodies

or antibody (FAB)-fragments directed against tumour associated antigens /46, 110, 122/. *Application:* in radioimmunoscintigraphy; experience in different tumours (colon, ovarian, breast carcinoma) have been published. Trials for application to radioimmunotherapy are in progress.

^{131}I-(^{125}I)-ligands for estrogen-receptor determination /35, 160/

In the last few years a great variety of estrogen-derivates has been labelled with ^{125}I, ^{18}F, ^{77}Br, ^{11}C for differentiation of estrogen receptor containing (or not containing) human breast carcinoma with respect to the selection of a suitable form (hormonal or chemotherapeutic) treatment for metastases. Tumours devoid of estrogen receptor are not responsive to hormonal treatment, contrary to the stronger response to receptor containing tumour cells. Application also to scintigraphy of receptor positive tumour metastases /49, 123/.

^{123}I-amphetamine (Figure 4.5.)

N-isopropyl (^{123}I)-p-iodoamphetamine (IMP) is commercially available, passes through the blood brain barrier attributable to free diffusion of the unionized lipophilic form of the compound, is taken up in the brain cells /154/. The initial brain uptake (1 h) reflects cerebral blood flow: a rapid washout of IMP-metabolites from grey matter may be due to its higher blood flow. The highest brain uptake (higher concentrations in grey matter, lower concentrations in white matter) in a similar kinetics to IMP was found with N,N,N'-trimethyl-N'-[2-hydroxy-3-methyl-5-iodobenzyl]-1,3-propane diamine (*HIPDM*) /81/. These are applied in SPECT-brain-scintigraphy to detect regional blood-perfusion disturbances and functional lesions, to differentiate partial epilepsy, and in patients with normal pressure hydrocephalus.

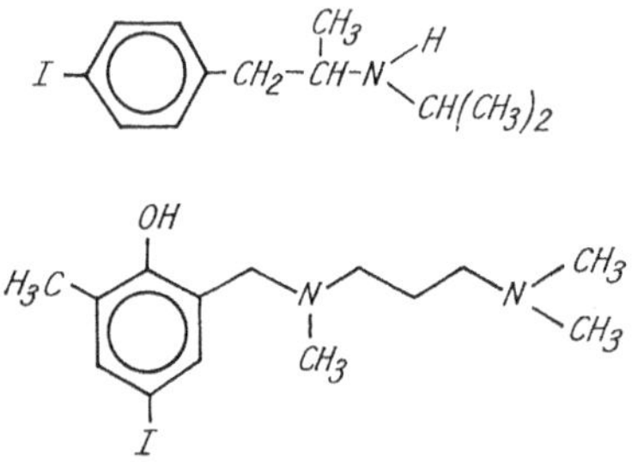

Fig. 4.5. Chemical structure of IMP and HIPDM

^{123}I-fatty acids

Labelling techniques with ^{131}I-monochloride to the double bound led to changes in physiological properties. Labelling with ^{123}I by interhalogen replacement on the terminal atom did not change the biological behaviour /106, 112/. With 16-^{123}I-hexadecanoic acid myocardial uptake and flow measurements were investigated (diagnostics in myocardial infarction).

3. Technetium-radiopharmaceuticals

Principles of ^{99m}Tc-labelling methods and all the problems concerning the reduction from Tc + 7 (in aqueous solution the most stable from) to + 3, + 4, + 5 and the mechanism of complexation of reduced Tc with pharmaceuticals will be disussed in Chapters 7 and 8.

^{99m}Tc-pertechnetate ($^{99m}TcO_4^-$)

The eluate of the ^{99}Mo/^{99m}Tc-generator usually with saline is pertechnetate. Tc has a shelf-life of 6 hours, a monoenergetic γ-energy: 0.140 MeV. It may be administered orally or intravenously. In the blood stream it is bound to transferrin, then it will be concentrated in the thyroid, chorioid plexus, mucosa of stomach and colon, and excreted via salivary glands and kidneys and retained in the bladder /3, 32, 62, 63, 65, 92/.

Application: Thyroid uptake (instead of radioiodine uptake test) in connection with the suppression test, thyroid scintigraphy, functional scintigraphy of salivary glands, brain scanning, blood-pool scintigraphy.

In all indications except thyroid and salivary diagnostics the patient has to receive potassium perchlorate (0.5 mg) per os or i.v. to block ^{99m}Tc-uptake in different tissues and lower radiation dose.

^{99m}Tc-labelled compounds

Drug regulations in various countries allow the use of only commercially available and controlled labelling kits for the preparation of ^{99m}Tc-radiopharmaceuticals. These kits contain the reducing agent (mostly stannous chloride, ascorbic acid freezedried in lyophilized form. Nowadays *single-step kits* are usually prepared: The $^{99m}TcO_4^-$-eluate has to be put into the vial and after stirring gently for a few minutes the freezedried substance is solved and the stable complex is built with a yield of 99 %. To remove the excess of pertechnetate is only necessary, if it is used for quantitative measurements by passing the solution through an anion-exchange resin. The shelf-life of the kits is up to six months after manufacture. The period of possible usage is printed on the cover.

Tc with the valence state + 7 ($^{99m}Tc^{+7}$) will be reduced by $SnCl_2 \cdot 2\,H_2O$ to valence state $^{99m}Tc^{+4}$. From this state the colloidal $TcO(OH)^{+2}$ is formed by hydrolysis. Chelating agents (EDTA, DTPA) are complexing this cation into a stable form and are ingredients of labelling kits.

^{99m}Tc-complexes are sensitive to oxidation (atmosphere, oxidizing agents) resulting in a breakdown of the complex. This can be avoided by preparation or storage in nitrogen atmosphere.

^{99m}Tc-human serum albumin (HSA)

Different reducing agents as iron ascorbic acid, stannous chloride or electrolysis have been used for ^{99m}Tc-binding to HSA /11, 36, 40, 70, 71, 78, 93/.

Application: measurement of blood volume and cardiac output, blood pool imaging (heart, placenta etc.), nowadays mostly replaced by labelled RBC, as HSA-aerosol inhalation for examining mucociliary clearance. Commercially available ^{99m}Tc-HSA-colloids with a particle size of 0.08 μm are used for lymph- and bone marrow scintigraphy.

^{99m}Tc-HSA macroaggregates and microspheres
Lyophilized macroaggregates (MAA) just like microsphere (HAM) may be solubilized in a stannous chloride solution; reduced Tc binds to the particles (preparation see Chapter 4, 23). Some prescriptions recommend to sonicate the suspension (5 min) to avoid an aggregation of particles, which may be built during standing after the labelling process /11, 71, 78/.
Application: Lung perfusion scintigraphy (MAA); microspheres have a greater spectrum of application according to the exactly defined diameter of particle size: ~ 80 nm for bone marrow scanning, 0.5—1 μm for liver, spleen-RHS-scintigraphy and functional examination of the RHS ("milli-microspheres"); 10—40 μm for lung scintigraphy.

^{99m}Tc-Fibrinogen
Labelling kit with lyophilized $SnCl_2 \cdot 2\,H_2O$, buffer (pH 9.5) and human fibrinogen prepared from hepatitis B surface antigen (HBsAg)-free plasma. Application to scintigraphy of deep vein thrombosis in statu nascendi /75/.

^{99m}Tc-Sulphur-colloid, ^{99m}Tc-Antimony-Sulphur-colloid
Various different labelling prescriptions exist with different results in particle size and surface. The kits contain sodium thiosulphate and an acid, gelatine Polyvinylpyrolidone as a stabilizer for the prevention of aggregating and growing into larger particles and EDTA. This is chelating Al^{3+}-ions which may be present in the eluate as a break-through of the generator column interfering with colloid formation /35, 64/.
Different colloid-building preparations are available: Sulphur colloid, antimone trisulphide colloid, rhenium sulphide colloid. *Application:* scintigraphy of RHS (liver, spleen, bone marrow). Localization of gastrointestinal bleeding. Small size-preparation (0.05 μm diameter) were used for lymph node scanning. Nebulized Tc-S-colloid (particle size 0.1—0.3 μm) are used for aerosol-inhalation lung scintigraphy (ventilation studies; pulmonary, epithelial permeability examinations).

^{99m}Tc-phytate
Phytate = inositolhexaphosphoric acid ester. Kits are available for in vivo-labelling: phytate sodium salt solution, which is intravenously injected, forms insoluble phytate colloid in reaction with plasma calcium ions. In vitro-preparation: phytate and calcium chloride are building the colloid during the labelling process /131/. *Application:* Liver-spleen-RHS-scintigraphy.

^{99m}Tc-DTPA (Figure 4.6.)
Diethylenetriaminepentaacetic acid (DTPA) is usually complexed with Tc, which is reduced in a stannous chloride solution. Different kits are available with $SnCl_2$ (colourless clear solution) or with ferrous ascorbate (greenish-yellow solution) as reducing agent. Tc-DTPA is nearly not bound to plasma proteins, not metabolized and excreted by glomerular filtration without tubular reabsorption; it has a renal extraction efficiency of 20—25 % /42, 67/.
Application: clearance studies (glomerular filtration rate), kidney sequential scintigraphy, brain scanning, cisternoscintigraphy, nebulized as aerosol for lung inhalation (ventilation) scanning.

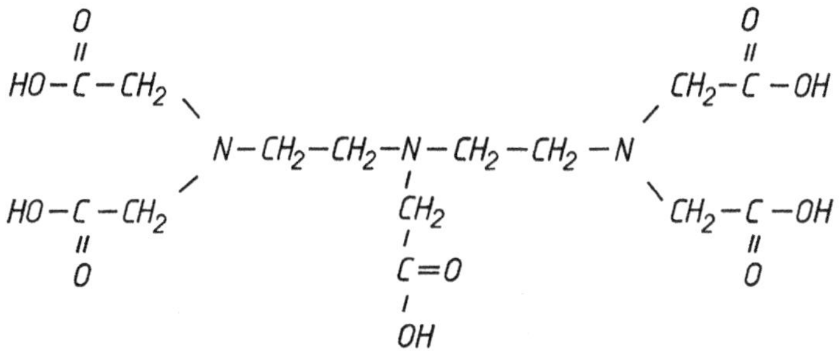

Fig. 4.6. Diethylenetriaminepentacetic acid (DTPA)

^{99m}Tc-glucoheptanoic acid (GHA; Figure 4.7.)
Labelled with a commercially kit /20, 151/ *Application:* brain scintigraphy (superior to Tc-DTPA or Tc-pertechnetate) and for dynamic kidney imaging.

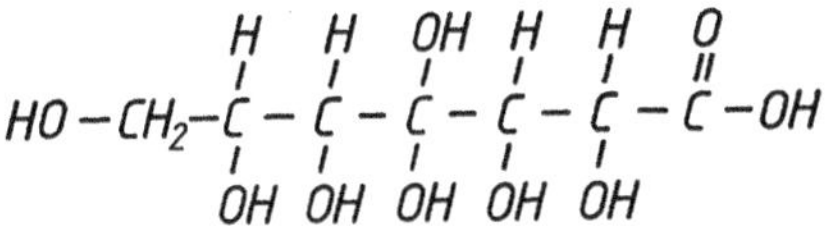

Fig. 4.7. Glucoheptanoic acid (GHA)

^{99m}Tc-DMSA (Figure 4.8.)
Dimercaptosuccinic acid (DMSA) is a sulphur containing chelating agent which complexes reduced ^{99m}Tc. This kit-prepared substance (lyophilisate of DMSA, $SnCl_2 \cdot 2\,H_2O$, ascorbic acid, pH 3.5—5.0) is nowadays the most convenient radiopharmaceutical for static kidney imaging with a higher concentration in the kidney parenchyma (60—70 % of test dose) and a longer retention compared to GHA /85/. It is largely plasma protein-bound and shows uptake in proximal tubules and Henle's loops and a slow urinary excretion. It can be used to detect proximal tubular function. Tc-DMSA has replaced ^{197}Hg-chlormerodrine and ^{203}Hg-Mersalyl used in the past.

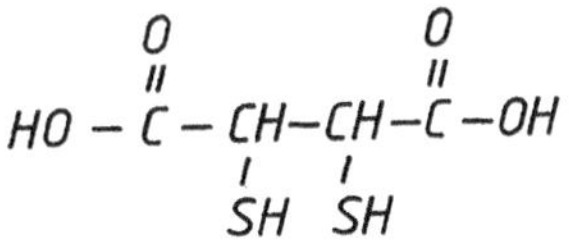

Fig. 4.8. Dimercaptosuccinic acid (DMSA)

^{99m}Tc-IDA (Figure 4.9.)
Several derivatives of the iminodiacetic acid (IDA) have been labelled with reduced Tc, three (HIDA, PIPIDA, butyl-IDA) are commercially available as labelling kits for Tc-IDA-complexation /66, 87, 143/. These were used as

a tracer for imaging the biliary excretion and the biliary system including gall-bladder function. Rapid and high liver uptake and excretion is desired for clear delineation of bile ducts. Substitutions on the benzene ring result in increasing molecular weight and lipid solubility. IDA derivatives have the same liver cell excretion mechanism (anion-clearance) as bilirubin; a pathologic high bilirubin level in serum leads to a competitive inhibition of IDA-excretion. Compounds are desired which show only little compensatory renal excretion: in elevated bilirubin serum level butyl-IDA shows the best

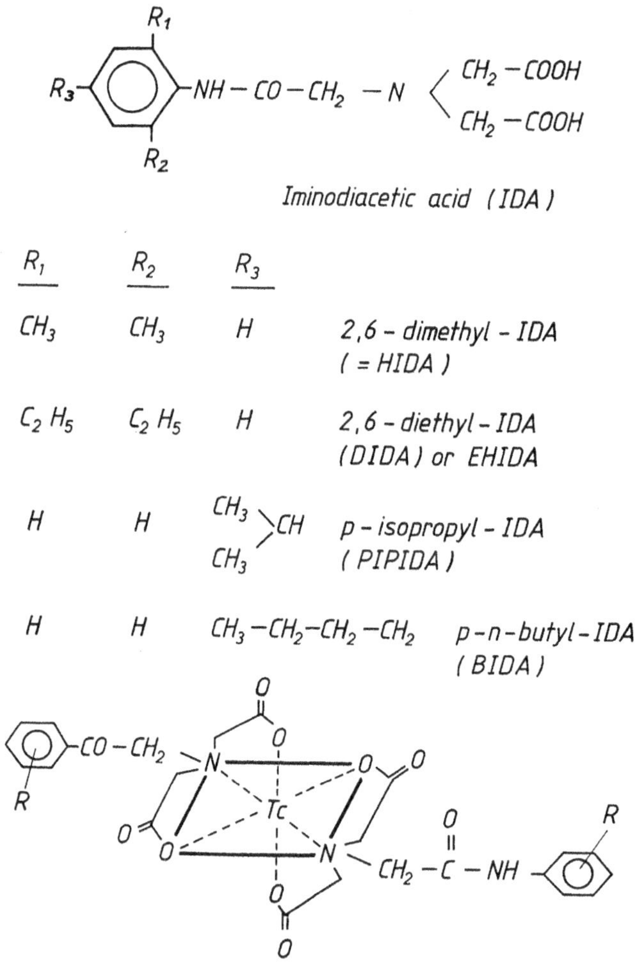

Fig. 4.9. Tc(III)-complex with acetanilidoderivatives of N-substituted iminodiacetic acid

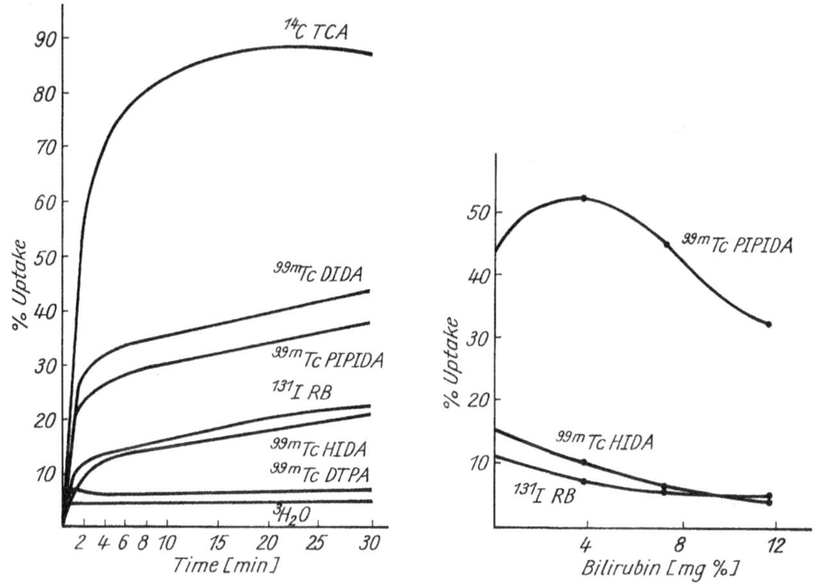

Fig. 4.10.a. Different accumulation of cholescintigraphic agents by isolated hepatocytes. Taurocholate (TCA), diethyl-IDA (DIDA), p-isopropyl-IDA (PIPIDA), rose bengal (RB), dimethyl-IDA (HIDA), diethylene triamine penta acetic acid (DTPA) as reference compound (Loberg and Porter, Soc. Nucl. Med., Inc. 1979, Nr. 79 — 67,730)

Fig. 4.10.b. Differences of uptake of cholescintigraphic agents ^{99}Tc-PIPIDA and -HIDA, compared to ^{131}I-rose bengal as a function of serum bilirubin level (Loberg and Porter, Soc. Nucl. Med., Inc. 1979, Nr. 79 — 67,730)

liver uptake and bile system delineation (Figure 4.10.). *Application:* cholescintigraphy, differentiation of functional disturbance of gall-bladder, differentiation of obstructive from parenchyma induced jaundice, detection of biliary reflux into the stomach, follow-up of surgical biliodigestive anastomosis.

^{99m}Tc-phosphate compounds (Figure 4.11., 4.12.)
Subramanian and *McAffee* introduced ^{99m}Tc-polyphosphate in 1971 for bone scintigraphy /129/. Polyphosphate and different phosphonates (dicarboxypropan-diphosphonic acid (DPD), methylenediphosphonate (MDP)) have been proposed for the same purpose /22, 104, 130, 132/. Their different kinetic behaviour depends on the various P-O-P and P-N-P-linkage in their molecule. Rapid blood clearance, high and specific skeleton uptake and rapid renal excretion with optimal bone/soft tissue ratio are desired. Best results were observed with ^{99m}Tc-HMDP; this is even superior to the MDP nowadays used as radiopharmaceutical of choice. Even higher skeleton uptake was found in the substances ^{99m}Tc-3,3-diphosphono-1,2-propanedicar-

boxylic acid (DPD) and ^{99m}Tc-hydroxymethylene diphosphonate (HDP), which are until now not commercially available /13, 21/.
The ^{99m}Tc-phosphate complexes show now plasma protein binding and a renal excretion by glomerular filtration without tubular reabsorption. During biokinetics these complexes were partially hydrolized ($TcO_2 \cdot 2\,H_2O$; Tc (4)) in different intensity depending on the complex stability.

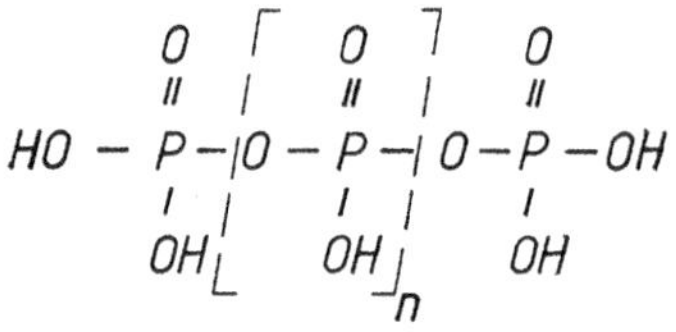

Polyphosphate

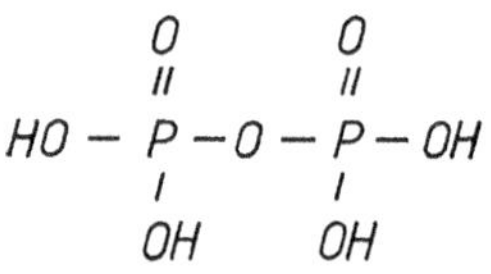

Pyrophosphate

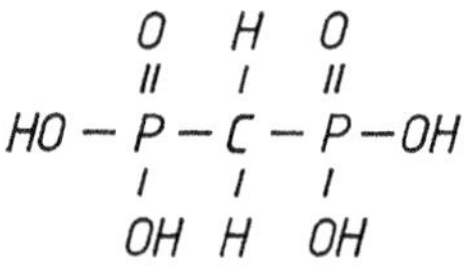

Methylene diphosphonate (MDP)

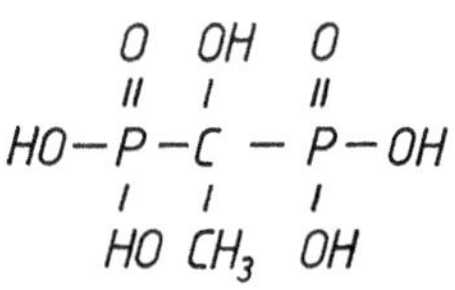

Ethane-hydroxy-diphosphonate (etidronate, EHDP)

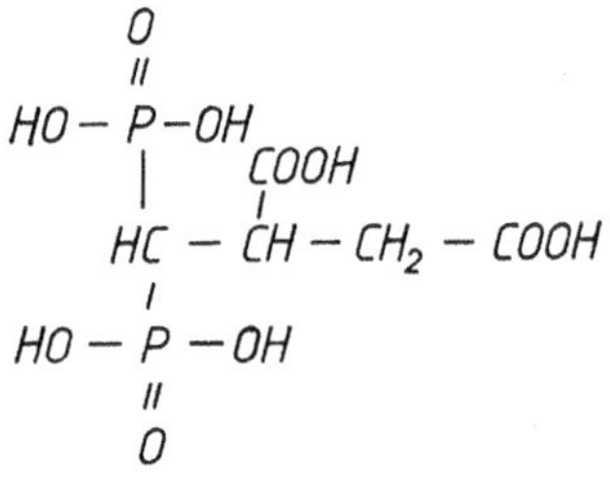

2,3 Dicarboxypropane-1,1 Diphosphonate (DPD)

Fig. 4.11.

These microcolloids were accumulated in kidney-parenchyma, in excudates, in tissue necrosis; they were plasma protein bound and show a tubular reabsorption. *Application:* bone scintigraphy (accumulation in areas with high metabolism as osteomyelitis, aseptic necrosis, bone fracture, bone resection and implantation in endoprothesis, tumour and metastases), scintigraphy of joints (arthritis, periarthritis), detection of calcification processes (myositis ossificans), localization of fresh myocardial infarction (^{99m}Tc-pyrophosphate is favoured).

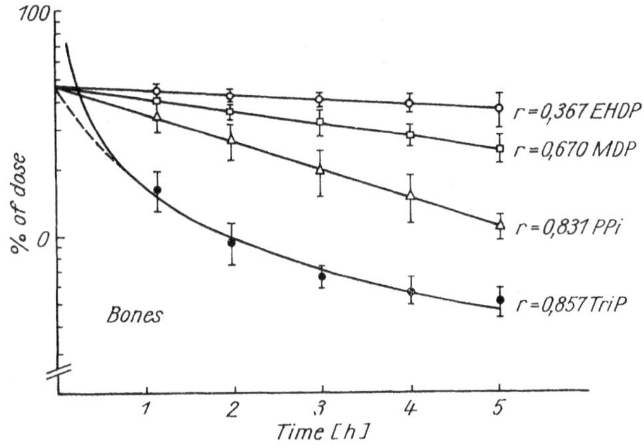

Fig. 4.12. Differences in bone retention time of different bone seeking ^{99m}Tc-labelled compounds: EHDP = hydroxy-ethanediphosphonate, MDP = methylene diphosphonate, PPI = pyrophosphate, TriP = triplephosphate. (Schuemichen et al. J. Nucl. Med. **21** (1980), 1080)

Further ^{99m}Tc-compounds

The labelled radiopharmaceuticals with only limited use are *^{99m}Tc-tetracycline* (scintigraphy of kidneys, gall-bladder, tumour, myocardial infarction); *^{99m}Tc-citrate* and ^{99m}Tc-thioglucose /34/ for kidney imaging; *^{99m}Tc-gelatine* /107/ (kidney and liver scanning); *^{99m}Tc-penicillamine* (gall-bladder- and cholescintigraphy /60, 73, 77, 146/), *^{99m}Tc-pyrilidoxiliden-glutamate* /5/ (cholescintigraphy), ^{99m}Tc-labelled *mercaptoisobutyric acid* (MIBA) /86/, *-toluidineblue* /159/, *-bilirubin* /75/ for liver function scintigraphy, *^{99m}Tc-antimone sulphidecolloid* for liver-RHS-scintigraphy and imaging of lymphatic system (colloid droplets of 0.03—0.3 μm diameter), *^{99m}Tc-urokinase* and ^{99m}Tc-streptokinase /105, 111/ (thromboscintigraphy) /45/, *^{99m}Tc-plasmin* /46, 102/ (a natural thrombolytic enzyme) is used as plasmin-uptake test in detecting deep vein thrombosis, ^{99m}Tc-labelled leucocytes for the detection of infection-lesions and abscess.

^{99m}Tc-erythrocytes

Labelling can be performed in vitro as well as in vivo: autologous red blood cells (RBCS) are withdrawn from patient, possibly separated from plasma,

incubated with Sn^{2+} (Sn-pyrophosphate, Sn-DTPA), which enters the cell, then $^{99m}TcO_4^-$ is added and will be reduced and bound relatively strictly to hemoglobin with a yield of $> 95\,\%$. Some authors use instead of the pretinning of RBC the posttinning techniques: stannous ions are added after Tc-eluate. In vivo-method: intravenous injection of stannous pyrophosphate or DTPA from a labelling kit, 20 min later injection of pertechnetate. In a combination of these two methods 20 min after Sn $^{2+}$-injection blood is withdrawn in a syringe containing pertechnetate. Reinjection is performed after 10–15 min of incubation in the syringe /12, 48/. *Application:* blood pool-scintigraphy, especially ventriculoscintigraphy with calculation of cardiac dynamic parameters (cardiac output, ejection fraction etc.), determination of blood volume, cardiac shunt-diagnosis, peripheral arteriography, localization of gastrointestinal bleeding. Heat damaged and denaturated RBCs are used for the determination of spleen function and spleen scintigraphy.
RBC can be transformed into spherocytes by putting the vial with Tc-RBC in a water-bath (+ 49.5 °C) for 10 to 15 min.

^{99m}Tc-compounds in the making

^{99m}Tc-p-aminohippuric acid (PAH-analog) for renal function studies /17/. *^{99m}Tc-DADS* (N,N'-bis-(mercaptoacetyl)-ethylenediamine) and *$^{99m}Tc\text{-}CO_2$-DADS* (N,N'-bis-mercaptoacetyl-2,3-diaminopropionate) were developed in order to replace ^{131}I-hippurate for renal function examination /18/. They were tubularly secreted; the kidney background ratio is 50 % higher relative to ^{131}I-hippurate /77, 81, 148/. See also p. 86.
Deutsch et al. were the first who tried in 1981 to replace ^{201}Tl-chloride as myocardial agent by cationic ^{99m}Tc-complexes: *^{99m}Tc-Tris-DMPE* (tris-bis [1,2-dimethylphosphino]ethane) produced the best results of all newly proposed ^{99m}Tc-compounds /37, 99/. New developments are taking place with the aim to find lipid-soluble Tc-labelled compounds, which are crossing the blood brain barrier for detection of perfusion and cell function in the central nervous system. Ligands based on bisaminoethanethiol (BAT)-chelating groups showed in vivo-stability and the ability to cross blood barrier /25, 50, 79/. ^{99m}Tc-labeled *propylene amine oxim (PnAO)* see p. 86 and 115.

4. Indium-111-, Indium-113m-compounds

The cyclotron-produced ^{111}In has ideal characteristics (Tp 67.4 h); γ-energy: 173 MeV (89 %); 247 MeV (94 %) for scintigraphy; its longer shelf-life compared to ^{99m}Tc and ^{113m}In is useful for functional studies over a period of some days depending on the pharmacokinetics of various compounds especially in inflammation and oncologic examinations. In $^{3+}$-ions form stable complexes with the chelates EDTA, DTPA and transferrin (see Chapter 9). It occurs in oxidation states + 1, + 2, + 3; the In^{3+} ions tend to hydrolize and build colloids (*In-hydroxide; *In-phosphate) /51, 52, 109, 126, 128/. ^{111}In is cyclotronproduced, available as trichloride in 0.05 n HCl /140/. ^{113m}In is generator ($^{113}Sn/^{113m}In$)-produced and eluted with 0.05 m hydrochloric acid (Tp: 104 min; γ-energy: 0.393 (64 %); the parent ^{113}Sn decays with a half-life of 115 d with γ 0.255 (1.8 %) being emitted, X-rays and electron capture) /37/. *In, which is injected intravenously, is strongly bound to plasma transferrin, used for blood pool diagnostics and circulation studies. Dissociating *In-complexes release the metal and this forms an *In-transferrin complex /127/.

^{111}In-EDTA, ^{111}In-DTPA were used for cisternography, detection of cerebrospinal fistula with less adverse reactions than ^{131}I-HSA and for clearance studies (glomerular filtration rate) /28/.

***In-transferrin** can be labelled in vivo (by injecting $InCl_3$) or in vitro. It is retained in the blood compartment for a long time and used for circulation studies, placentography, bone marrow and tumour scanning /53, 72, 127/. With the help of a covalent bifunctional metal chelate (azo-EDTA) *In may be bound to proteins by means of a diazobound: *^{111}In-fibrinogen* for thrombosis detection in statu nascendi; *^{111}In-HSA* is used for circulation studies and *^{111}In-bleomycine* as an unspecific positive tumour-scanning agent (see ^{57}Co-bleomycine) /138, 147/.

^{111}In-monoclonal (polyclonal) antibodies or Fab-, F(ab')$_2$-fragments against different specific tumours have been developed during the last few years.

^{111}In-oxine (^{111}In-8-hydroxyquinoline-complex) is used for labelling blood cells (leucocytes, lymphocytes, platelets) without loss of their cell-function: oxine used as ^{111}In-carrier is lipid soluble and passes the cell membrane. ^{111}In-RBC-labelling is a seldomly applied alternative to ^{99m}Tc-erythrocytes /100/ ^{111}In-platelets can be applied to non-invasively kinetic and survival studies as well as for the detection of deep vein thrombosis, intraventricular and coronary thrombi and acute renal transplant rejection /88, 139/.

^{111}In-leucocytes were applied to granulocytokinetics and to localizing inflammation or abscess-scintigraphy /88, 141/.

5. Thallium-201

The cyclotron-produced ^{201}Tl decays to ^{201}Hg with Tp 74 h and γ-energy: 0.135 (2 %) and 0.167 MeV (8 %); X-rays: 69—83 MeV (93 %), electron capture. It is available as ^{201}Tl-chloride in physiologic saline. After rapid blood clearance it is accumulated in the myocardium, thyroid, kidneys and large bowel /83/. *Application:* myocard scintigraphy under stress and hours later in redistribution for the detection of infarcted areas and differentiation from stress-induced ischemia. Thallium acts like potassium in the myocardial uptake according to coronary perfusion. In research one tries to replace ^{201}Tl, because of its high costs, by ^{99m}Tc-compounds with similar biological behaviour. Further indications are thyroid tumour scintigraphy (with differing results) and the detection of parathyroid adenomas in connection with pertechnetate substraction scanning.

6. Hg-203- and Hg-197-compounds

^{203}Hg: Tp 46.9 d; γ-energy: 0.279 (77 %); β-energy: 0.212 MeV (100 %), ^{197}Hg: Tp 64.4 h; γ-energies: 0.077 and 0.067 MeV; e c; no β-rays. In the past, mercury-compounds were used for static kidney imaging as ^{197}Hg-chlormerodrine (Neohydrin®) and ^{203}Hg-mersalyl (Salyrgan®) with a long-term retention in the proximal renal tubules. In the past it was used for brain scintigraphy and gammaencephalography /91, 121/. ^{197}Hg-brommercuryhydroxypropane (BMHP) was applied in spleen-scintigraphy in the past. Because of the high radiation dose all mercury-agents are nowadays worldwide replaced by ^{99m}Tc-agents /31, 149/.

7. Au-198-colloid

^{198}Au: Tp 2.7 days; γ-energy: 0.412 MeV (95 %); β-energy: 0.962 MeV (100 %) (Table 4.2.). The deep cherry-red solution colloidal gold contains reducing agents (ascorbic acid) and gelatine as colloid-stabilizer. It is commercially available and was used for a long time as a radiopharmaceutical of choice for liver-RHS-scintigraphy and liver-function studies (particle size 30 to 50 nm) and for lymphoscintigraphy (2—10 nm). Nowadays these products are replaced by ^{99m}Tc-compounds, though radiogold colloids are more stable in vivo and have a relatively more homogenous particle distribution than ^{99m}Tc-S-colloid. ^{198}Au-colloid is nowadays used for intracavitary treatment in malignant peritoneal or pleural effusions and for radiosynovectomy of knee-joints /4, 95, 96/. Intrathecally applied it is used as prophylactic treatment of meningosis leucaemica. ^{198}Au-metal is used as wire for implantation treatment /118, 119/.

8. Rhenium-186

^{186}Re: Tp 3.68 days; β^- 1.07 MeV (95 %); γ-energy: 0.137 MeV (9 %). ^{186}Re-colloid contains ascorbic acid and gelatine as stabilizers. It is used for radiosynovectomy of larger joints (elbow, knee). In its physical characteristics it is comparable to ^{198}Au-colloid (Table 4.2.).

9. Erbium-169

^{169}Er: Tp 9.6 days, γ-energy: 0.008 MeV (0.39 %), β^-: 0.34 MeV (100 %). ^{169}Er-citrate or ^{169}Er-iron-colloid is used as colloid solution for radiosynovectomy of small joints (wrist, finger) because of its low β-energy (Table 4.2.).

Table 4.2. Radiopharmaceuticals for treatment

Radio-nuclide	Compound	half-life (Tp)	γ-energy (MeV)	β-energy (MeV)	Penetration in tissue (mm) maxi-mum	mean	Indication
^{32}P	sodium phosphate	14.5 d		1.71 max.	8.0	1.9	polycythaemia vera, bone metastasis
	chromic phosphate colloid			0.7 mean			intracavitary treatment
	tri-sodium octylphosphate						endolymphatic treatment
^{89}Sr	chloride						bone metastasis
^{90}Y	silicate	2.7 d		2.27 max.	11.0	3.6	intracavitary treatment
	ironhydroxide colloid			0.9 mean			radiosynovectomy (large joints)
^{125}I	sodium iodide	60.2 d	0.035 X-ray, EC, K				hyperthyroidism
^{131}I	sodium iodide	8.0 d	0.364	0.61/0.81	2.2	1.5	hyperthyroidism, thyroid adenoma, euthyroid goiter, postsurgical recurrence of goiter, thyroid carcinoma metastasis, total thyroidectomy remnants
^{169}Er	citrate colloid	9.4 d	0.008	0.34 max.	1.0	0.3	radiosynovectomy (small joints)
^{186}Re	sulphide colloid	90 h	0.137	1.07 max.	4.0	1.2	radiosynovectomy (large joints)
^{198}Au	colloid	2.7 d	0.412	0.96 max. 0.32 mean	4.0	1.2	intracavitary treatment, radiosynovectomy, intrathecal prophylaxis of meningosis leucaemica

10. Ytterbium-169

^{169}Yb: Tp 32 days; γ-energy: 0.063 MeV (45 %), 0.107 (22 %); 0.198 (35 %) applied as ^{169}Yb-DTPA to cisterno- and myelography and for clearance examinations (glomerular filtration rate) /73/.

11. Xenon-133

^{133}Xe: Tp 5.4 days; γ-energy: 0.081 (37 %); β: 0.346 MeV; X-ray and K-radiation. Radioxenon is a water-insoluble noble gas, available in ampoules or as a solution dissolved in sterile isotonic saline. After intravenous injection it is used for regional lung perfusion and exhalation scintigraphy, for cardiac shunt detection and the measurement of muscle blood flow. As a gas it is applied in regional lung ventilation scintigraphy. Applied by intraarterial catheter, organ perfusion determinations of the liver, kidney or brain can be performed /6, 80/.
Nowadays it can be replaced by *Krypton-81m* delivered from ^{81}Rb/^{81m}Kr-generator, which is expensive and usable only for a short time /158/. The parent ^{81}Rb decays with a half-life of 4.6 h with emission of γ-rays, β^+ and EC to the metastable ^{81m}Kr (Tp: 13 s, γ-energy: 0.191 MeV (65 %)). *Application:* cardiac and circulatory studies with krypton-eluate, lung perfusion and ventilation studies (with krypton gas).

12. Yttrium-90

^{90}Y: Tp 2.66 days; β-energy: 2.27 MeV (100 %), no γ. ^{90}Y-citrate or ^{90}Y-iron hydroxide-colloid or ^{90}Y-Fe-silicate is used for treatment, when relatively high β-energy is used: radiosynovectomy of knees and intracavitary treatment (peritoneal, pleura tumor effisions; brain tumour after partial resection) /112/ (Table 4.2.).

13. Strontium-87m

^{87m}Sr can be eluted from a ^{87}Y/^{87m}Sr-generator. The parent nuclide decays with a half-life of 80.3 h, with γ-, β-rays being emitted and EC into ^{87m}Sr: Tp 2.8 h, γ-energy: 0.388 MeV (80 %). It was used as ^{87m}Sr-chloride for bone scintigraphy (rapid blood clearance, high skeleton accretion), nowadays replaced by ^{99m}Tc-phosphonate compounds /2, 23, 125/.

14. Strontium-85

^{85}Sr (Tp 64.0 d, γ-energy: 0.510 MeV) used as ^{85}Sr-citrate or -chloride was used for bone-profilography in the past /8, 9, 123, 124/.

15. Strontium-89

^{89}Sr (Tp 52.7 days, γ-energy: 0.91 MeV (0.009 %), β^--energy: 1.463 MeV (100 %)) can be applied to palliative treatment of bone metastases to reduce pain.

16. Selenium-75-compounds

^{75}Se: Tp 120 days; γ-energy: 0.136 MeV (57 %), 0.265 MeV (60 %).
^{75}Se-Selenite was used as an unspecific tumour agent for scintigraphy /47/. In organic compounds ^{75}Se replaces Sulphur:

Selenium-75-Selenemethionine

^{75}Se-Selenemethionine was abandoned because of its high radiation load /15/.
It was applied for a long time as a radiopharmaceutical of choice in pancreas and parathyroid scintigraphy; physiologically it is retained in the liver, so liver-colloid-scintigraphy had to be subtracted from liver-pancreas-scan.

Selenium-75-nor-cholesterol

^{75}Se-nor-cholesterol (Scintadren®, see Chapter 18) is the only ^{75}Se-compound used in clinical practice (see p. 89), applied in adrenal cortex-scintigraphy /24/.
Experimental trials with ^{75}Se-tertiary amines have been published, which are lipid-soluble, pass the blood brain barrier and were retained in brain cells (see p. 85). Their replacement by more suitable labels is now being researched.

17. Gallium-67

^{67}Ga (Tp 78 h), γ-energy: 0.093 MeV (40 %); 0.184 MeV (24 %); 0.296 MeV (22 %). $^{67}Ga^{3+}$ is cyclotron-produced and as ^{67}Ga-citrate available, i.v. injected it is bound and complexed to plasma transferrin; furthermore, it is distributed and accumulated in the liver, spleen, lacrimal glands, nasopharynx and skeleton. Imaging is not possible before 3 days after injection: cleaning of digestive tract (laxatives) or repeated scanning is recommended to avoid false-positive focal accumulations /44/. *Application:* tumour scintigraphy; it accumulates in several tumours, in lymphomas (malignant lymphomatosis), in inflammation processes (abscess-localization) in sarcoidosis. There are several mechanisms of uptake in inflammatory lesions:

a) uptake as a result of higher blood supply and increased capillary permeability,
b) uptake in granulocytes, bacteria and bacterial siderophores,
c) uptake in lactoferrin,
d) uptake in reactive bone (osteomyelitis).

Mechanisms of uptake into tumour cells are discussed: diffusion into tumour cells, binding to lysozymes, lysosome-like granula, binding of ionic $^{67}Ga^{3+}$ by tumour proteins, exchange with intracellular calcium. The uptake is increased more in poorly differentiated tumours and especially in lung-, breast-cancer and osteocarcinoma /8/. The generator (^{68}Ge/^{68}Ga)-produced

18. Gallium-68

^{68}Ga is used as *^{68}Ga-EDTA, ^{68}Ga-transferrin, ^{68}Ga-microspheres,* and as a label for blood cells (see Chapter 3). The parent nuclide Germanium-68 decays with a half-life of 288 days (EC) into ^{68}Ga (Tp 68 min; positron-emitter with annihilation γ-energy: 0.511 MeV) /89, 150/.

19. Copper-64

^{64}Cu, reactor-produced, Tp 12.9 h; γ-energy: 0.51 MeV from β^+; β^--energy: 0.573 MeV; EC, X-rays.
^{64}Cu-chloride or acetate is applied for Cu-kinetics in *M. Wilson:* i.v.-injected plasma concentrations are measured for 5 days and cumulative urine and faecal excretion for differential diagnosis (method of choice) of chronic liver disease and autosomal recessive copper metabolic disease (Morbus Wilson) with hepatolenticulare degeneration of CNS /153/.

20. Cobalt-compounds

^{57}Co (Tp 270 days; γ-energy: 0.122 MeV (85.2 %)
^{58}Co (Tp 70.8 days; γ-energy: 0.611 MeV (99.4 %)
^{57}Co is cyclotron-, ^{58}Co reactor-produced, available as *Co-chloride, used for biosynthesis of *Co-Vitamin B_{12}. This is applied orally to quantify its gas-

trointestinal resorption to detect Vitamin B_{12} deficiency as a cause of pernicious anemia or funicular medullosis /14, 30/.
**Co-bleomycine* is a chelate compound, in vivo-stable, with a complicated polypeptide structure. As an antimitotic agent it has a high toxicity, the dose must be 2 mg. It can be used for tumour scintigraphy (liver, kidney, brain) with only slight effort /100/.

21. Ferrous-59-compounds

^{59}Fe, reactor-produced, Tp 44.6 days; γ-energy: 1.095 MeV (56 %) and 1.292 MeV (44 %); β-energy: 1.57 MeV.
The divalent ^{59}Fe is used as ^{59}Fe-ferrous citrate for ferrokinetic examinations (iron plasma clearance, iron turnover, iron-hemoglobin synthesis, enteral iron absorption, red bone marrow distribution). To preserve the divalent state the solution of ferrous citrate or sulphate contains reducing agents as an ascorbic or citric acid /19, 41, 74, 90, 108/.

22. Chromium-51-compounds

^{51}Cr, reactor-produced, Tp 27.8 days; γ-energy: 0.320 MeV (9 %), X-rays, EC.

Natrium-chromium-51-chromate

Na-^{51}Cr-chromate (valence state + 6) is used as a label for blood cells, especially platelets and RBC: it will be added to erythrocytes in ACD-solution (for stabilizing the 6 + valence state) and incubated 20 min at + 37 °C for binding Cr^{6+} into hemoglobin molecule. The addition of ascorbic acid reduces untagged Cr^{6+} to Cr^{3+}, which does not tag cells and is removed by plasma protein binding /1, 55/. After transformation of RBC into spherocytes by heat /155/ or chemical denaturation /31, 149/ it was used for spleen function studies or scintigraphy. *Application:* to erythrokinetics (survival studies, localization) of RBC-sequestration in the liver or spleen in hemolytic anaemias, to estimation of red cell mass and gut blood loss, to spleen scintigraphy with heat-damaged RBCs. ^{51}Cr as a cell label is more and more replaced by ^{99m}Tc, ^{113m}In or ^{111}In.

Chromium-51-EDTA

^{51}Cr-EDTA or DTPA is applied in the glomerular filtration rate determination in clearance examinations, partly replaced by ^{99m}Tc-, ^{113m}In- or ^{169}Yb-DTPA /33, 73, 124/.

Chromium-51-albumin

^{51}Cr-albumin is used for the determination of plasma protein turnover examinations and the detection of gastroenteral protein loss (exsudative gastroenteropathy) with whole-body counting or fecal excretion measurements.

23. Phosphor-32-compounds

^{32}P, reactor-produced, Tp: 14.3 days, pure β-emitter (1.71 MeV max, 0.7 mean). *^{32}P-diisopropyl-fluorophosphonate (DF^{32}P)* was used for in vivo labelling of platelets and red blood cells (RBC) /58, 84/.
^{32}P-orthophosphate was introduced into metabolic investigations by *George von Hevesy* and *Chievitz* in 1935 /26/. It is applied in eye tumour diagnostics as it accumulates in the DNA of rapidly growing tissue in the ^{-32}P-test performed with semiconductor-detector contact measurements of the bulb. Further application in the intraoperative detection of tumours (brain, parathyroid, adrenal). As a pure β-emitter (see Table 4.2.) ^{32}P was introduced for treatment of leucemia by *Lawrence* et al. in 1939 /61, 82/. Today it is applied for treatment of polycythemia vera, for intracavitary treatment as *^{32}P-chromic phosphate colloid,* as *^{32}P-triocyle-phosphate* for the treatment of the lymphatic system and as *^{32}P-orthophosphate* for palliative therapy of pain in bone metastases /38, 43, 90/.

24. Fluorine-18

^{18}F is cyclotron-produced, Tp 2.8 h; γ-energy: 0.510 MeV from β^+ (0.635 MeV).
^{18}F-sodium fluoride or ^{18}F-aluminate complex had been used for bone scintigraphy in the past before ^{-99m}Tc-phosphonate complexes were introduced. It has optimal biokinetic characteristics (rapid blood clearance, high bone uptake and accretion by exchange of hydroxyl-groups in apatite) /16/.
^{18}F is used as a label for amino acids (pancreas scintigraphy with ^{18}F-tryptophance), for hormones (radioligands for estrogen-receptor studies), for sugars (^{18}F-amphetamines; ^{18}F-desoxyglucose-derivatives) and for fatty acids /136, 137/.

References: see Chapter 5.

Chapter 5

Historical Development of Radiopharmaceuticals: Bibliography

Harald Deckart

All the articles compiled below provide a *historical overview* of the development *of radiopharmaceuticals.* Most of them are original publications (first reports). References to Chapter 3 and 4.

1. *Aas, K. A., F. H. Gardner:* Survival of blood platelets labeled with chromium-51. J. Clin. Invest. **32** (1958), 1257.
2. *Allen, J. F., J. J. Pinajian:* A ^{87m}Sr-generator for medical applications. Int. J. appl. Radiat. **16** (1965), 319.
3. *Andros, G., P. V. Harper, K. A. Lathrop, R. J. McCardle:* Pertechnetate-99m localization in man with applications to thyroid scanning and the study of thyroid physiology. J. Clin. Endocrin. **25** (1965), 1067.
4. *Ansell, B. M., A. Cook, J. R. Mallard* et al.: Evaluation of intraarticular colloid gold, Au-198, in the treatment of persistent knee effusions. Am. Rheum. Dis. **22** (1963), 435.
5. *Baker, R. J., J. C. Bellen, P. M. Rohai:* Tc-99m-pyridoxylidene-glutamate: a new rapid cholescintigraphic agent. J. Nucl. Med. **15** (1974), 476.
6. *Ball, W. C. jr., P. B. Stewart, L. G. S. Newsham, D. V. Bates:* Regional pulmonary function studied with Xenon-133. J. Clin. Invest. **41** (1962), 519.
7. *Bauer, F. K.:* Myelography by means of ^{131}I. Neurology **3** (1953), 341.
8. *Bauer, G. C. H., R. D. Ray:* Kinetics of strontium metabolism in man. J. Bone Jt. Surg. **40** *A* (1958), 171.
9. *Bauer, G. C. H., B. Wendeburg:* External counting of ^{47}Ca and ^{85}Sr in studies of localized skeletal lesions in man. J. Bone Jt. Surg. **41** B (1959), 558.
10. *Beierwaltes, W. H., V. M. Varma, R. E. Counsell, L. H. Lieberman:* Scintillation scanning of malignant melanomas with radio-iodinated quinoline derivatives. J. Nucl. Med. **9** (1968), 489.
11. *Benjamin, P. P., A. Rajali, H. Friedell:* Electrolytic complexation of ^{99m}Tc at constant current: its applications in nuclear medicine. J. Nucl. Med. **11** (1970), 147.
12. *Berger, R., B. Johannsen:* Markierung von Erythrozyten mit Technetium-99m. Wiss. Z. KMU Leipzig, Math. Nat. Reihe 18 (1969), 535.
13. *Bevan, J. A., A. J. Tofe, J. J. Benedict, M. D. Francis, B. L. Barnett:* Tc-99m HMDP (Hydroxymethylene Diphosphonate): A radiopharmaceutical for skeletal and acute myocardial infarct imaging. I. Synthesis and distribution in animals. J. Nucl. Med. **21** (1980), 961.
14. *Blau, M.:* Biosynthesis of ^{75}Se-Selenomethionine and ^{75}Se-Selenocystine. Biochem. biophys. Acta **49** (1961), 389.
15. *Blau, M., R. F. Manke:* The pancreas specificity of ^{75}Se-selenomethionine. J. Nucl. Med. **2** (1961).
16. *Blau, M., W. Nagler, M. A. Bender:* Fluorine-18: A new isotope for bone scanning. J. Nucl. Med. **3** (1962), 332.
17. *Blau, M., R. R. Zielinski, M. A. Bender:* ^{137m}Ba cow—a new short lived isotope generator. Nucleonics **24** (1966), 60.

18. *Blaufox, M. D., D. R. Sanderson, W. N. Tauxe* et al.: Plasmatic disappearance of ^{131}I-diatrizoate and glomerular filtration in the dog. Amer. J. Physiol. **204** (1963), 536.
19. *Bothwell, T. H., A. V. Hurtado, D. M. Donohne, C. A. Finch:* Erythrokinetics IV. The plasma iron turnover as a measure of erythropoiesis. Blood **12** (1957), 409.
20. *Boyd, R. E., J. Robson, F. C. Hunt, P. J. Sorby, I. P. C. Murray, W. J. McKay:* ^{99m}Tc-Gluconate complexes for renal scintigraphy. Brit. J. Radiol. **46** (1973), 604.
21. *Buell, U., E. Kleinhans, E. Zorn-Bopp, W. Reuschel, W. Muenzing, E. A. Moser, M. Seiderer:* Comparison of bone imaging with Tc-99m DPD and Tc-99m MDP. Concise Communication. J. Nucl. Med. **23** (1982), 214.
22. *Castronovo, F. P., R. J. Callahan:* A new bone scanning agent: ^{99m}Tc-labelled 1-hydroxy-ethylidene-1,1-disodium phosphonate. J. Nucl. Med. **13** (1972), 823.
23. *Charkes, N. D., D. M. Sklaroff, J. Bierly:* Detection of metastatic cancer to bone by scintiscanning with strontium-87m. Amer. J. Roentgenol. **91** (1964), 1121.
24. *Chatal, J. F., B. Charbonnel, R. Gurhard:* Un nouveau radiotracer pour la scintigraphie des surrenales le 6-methyl-^{75}Se-selenocholesterol. Non Presse medi. **6** (1977), 1145.
25. *Chersu, L. R., B. M. Sundoro, M. D. Blaufox:* Technetium-99m-labelled p-aminohippuric acid analog: a new renal agent. J. Nucl. Med. **25** (1984), 1111.
26. *Chiewitz, O., G. von Hevesy:* Radioactive indicators in the study of phosphorus metabolism in rats. Nature (London) **136** (1935), 754.
27. *Ciscato, V. A., J. O. Nicolini, M. C. Palcos:* Albumin macroaggregates labeled with indium-113m for lung scintiscanning. Int. J. appl. Radiat. **20** (1969), 633.
28. *Clements, J. P., H. N. Wagner, H. S. Stern:* Indium-113m DTPA: a new radiopharmaceutical for brain scanning. Amer. J. Roentgenol. **104** (1968), 136.
29. *Counsell, R. E., V. V. Ranade, R. J. Blair, W. H. Beierwaltes, P. A. Winhold:* Tumor localizing agent. IX. Radioiodinated cholesterol. Steroids **16** (1970), 317.
30. *Cresseri, A., F. Marro:* Clearance renale della vitamina B_{12} nel coniglio. Boll. Soc. Ital. Biol. Sper. **33** (1957), 1662.
31. *Croll, M. N., L. W. Brady, I. Bradsky, L. Szanton:* A new agent for splenic scanning-BMHP. Radiology **84** (1965), 492.
32. *Deckart, H., M. Tautz:* Physikalische und biologische Studien mit Technetium-99m-Pertechnetat. Dt. Gesundh.-Wesen **21** (1966), 222.
33. *Deckart, H., H. Flentje, H. Herzmann:* ^{51}Cr-DTPA in der Nierendiagnostik. Nucl. Med. **7** (1968), 205.
34. *Deckart, H., J. Weiland, A. Blottner, M. L. Weiss:* ^{99m}Tc-Thioglukose – ein neues Radiopharmakon für die Nierendiagnostik. In: Radioaktive Isotope in Klinik und Forschung, Vol. 13/1. Egermann, Wien 1978, P. 97.
35. *Degrossi, O., J. S. Martinez, H. Gotta:* A new ^{99m}Tc-labelled colloid for liver scanning. Minerva nucl. **9** (1965), 424.
36. *Deutsch, M. E., M. L. Redmond:* Unitary freeze-dried kits for preparation of technetium labelled human serum albumin. J. Nucl. Med. **13** (1972), 426.
37. *Deutsch, E., K. A. Glavan, V. J. Sodd, H. Nishiyama, D. L. Ferguson, S. J. Lukes:* Cationic Tc-99m complexes as potential myocardial imaging agents. J. Nucl. Med. **22** (1981), 897.
38. *Dobson, E. L., H. B. Jones:* The behaviour of intravenous by injected particulate material. Acta Med. scand. **144** (Suppl. 273) (1952), 1.

39. *Doering, R. F., W. D. Tucker, L. G. Tang* jr.: A simple process for milking yttrium-90 from strontium-90. Rep. BNL-472 (1957).
40. *Dreyer, R., R. Muenze:* Zur Tc-99m-Markierung von Serumalbumin. Isotopenpraxis **5** (1969), 296.
41. *Dubach, R., S. T. E. Callender, C. V. Moore:* Studies in iron transportation and metabolism. Blood **3** (1948), 526.
42. *Eckelman, W. C., P. Richards:* Instant ^{99m}Tc-compounds. J. Nucl. Med. **10** (1969), 245.
43. *Edwards, J. M.:* Malignant melanoma: treatment by endolymphatic radioisotope infusion. Ann. R. Coll. Surg. Eng. **44** (1969), 237.
44. *Edwards, C. L., R. L. Hayes:* Tumorscanning with ^{67}Ga-citrate. J. Nucl. Med. **10** (1969), 103.
45. *Elias, H., C. H. Arnold, G. Kloss:* Preparation of ^{131}I-labelled m-iodohippuric acid and its behaviour in kidney function studies compared to o-iodohippuric acid. Int. J. appl. Radiat. **24** (1973), 463.
46. *Epenetos, A. A., S. Mather, M. Granowska, C. Nimmon, L. R. Hawkins, K. Britton, J. Shepherd, J. Taylor-Papadimitriou, H. Durbin, J. S. Malpas, W. F. Bodmer:* Targeting of iodine-123-labelled tumor associated monoclonal antibodies to ovarian, breast and gastrointestinal tumors. Lancet (1982), 999.
47. *Esteban, J., R. Lash, S. Perez-Modrego:* Detection of cartilagious tumors with Selenium-75. Radiology **85** (1965), 149.
48. *Fischer, J., R. Wolf, A. Leon:* Technetium-99m as a label for erythrocytes. J. Nucl. Med. **8** (1967), 229.
49. *Franke, L. A., R. N. Hanson:* Radioiodinated ligands, for the estrogen receptor. J. Nucl. Med. **25** (1984), 1116.
50. *Fritzberg, A. R., W. C. Klingensmith, W. P. Whitney et al.:* Chemical and biological studies of ^{99m}Tc-N,N-bis(mercapto-acetamido)ethylendiamine: a potential replacement for I-131-iodohippurate. J. Nucl. Med. **22** (1981), 258.
51. *Goodwin, D. A., H. S. Stern, H. N. Wagner* jr., *H. H. Kramer:* Indium-113m: A new radiopharmaceutical for liver scanning. Nucleonics **24** (1966), 65.
52. *Goodwin, D. A., P. Matin, R. A. Finston:* ^{111}In—A new radiopharmaceutical for extended studies. J. Nucl. Med. **11** (1970), 388.
53. *Goodwin, D. A., R. Goods, L. Brown* et al.: ^{111}In-labelled transferrin for the detection of tumors. Radiology **100** (1971), 175.
54. *Goris, M. L.:* ^{123}I-iodobromsulphalein as a liver and biliary scanning agent. J. Nucl. Med. **14** (1973), 820.
55. *Gray, S. J., K. Sterling:* The tagging of red cells and plasma proteins with radioactive chromium. J. Clin. Invest. **29** (1950), 1604.
56. *Greenwood, F. C., W. M. Hunter, J. S. Glover:* The preparation of ^{131}I-labeled human growth hormone of high specific radioactivity. Biochem. J. **89** (1963), 114.
57. *Gumpel, J. M., E. D. Williams, H. J. Glass:* Use of Yttrium-90 in persistent synovitis of the knee. Ann. Rheum. Dis. **32** (1973), 223.
58. *Hahn, L., G. von Hevesy:* A method of blood volume determination. Acta physiol. scand. **1** (1940), 3.
59. *Halpern, B. N., G. Biozzi, B. Benacerraf, C. Stiffel, B. Hillemand:* Cinetique de la phagocytose d'une serum albumine humaine specialment traitée et radiomarquee et son application à l'etude de la circulation hépatique chez l'homme. CR Soc. Biol. (Paris) **150** (1956), 1307.
60. *Halpern, S., M. Tubis, J. Endow, C. Walsh, J. Kunsa, B. Zwikker:* ^{99m}Tc-Penicillamine—a cetazolamide complex. A new renal scanning agent. J. Nucl. Med. **13** (1972), 45a.
61. *Hamilton, J. G., J. H. Lawrence:* Recent clinical developments in the thera-

peutic application of radiophosphorus and radioiodine. J. Clin. Invest. **21** (1942), 624.
62. *Harper, P. V., G. Autors, K. Lathrop:* Preliminary observations in the use of six hour Tc-99m as a tracer in biology and medicine. Argonne Cancer Research Hospital Semiannual Report to the Atomic Energy Commission **18** (1962), 76.
63. *Harper, P. V., G. Andros, K. A. Lathrop, W. Siemens, L. Weiss:* Technetium-99m as a biological tracer. J. Nucl. Med. **3** (1962), 209.
64. *Harper, P. V., K. A. Lathrop, P. Richards:* ^{99m}Tc as a radiocolloid. J. Nucl. Med. **5** (1964), 382.
65. *Harper, P. V., S. F. Mullan, R. Fink:* Rapid brain scanning with technetium-99m. Proc. Inst. Me. d Chic. **25** (1964), 131.
66. *Harvey, E., M. Loberg, M. Cooper:* ^{99m}Tc-HIDA: a new Radiopharmaceutical for hepato-biliary imaging. J. Nucl. Med. **16** (1975), 533d.
67. *Hauser, W., L. Atkins, K. G. Nielson, P Richards:* Technetium-99m-DTPA: a new radiopharmaceutical for brain and kidney scanning. Radiology **94** (1970), 679.
68. *Hertz, S., A. Roberts, R. D. Evans:* Radioactive iodine as an indicator in the study of thyroid physiology. Proc. soc. Exp. Biol. Med. **38** (1938), 510.
69. *Hertz, S., A. Roberts:* Application of radioactive iodine in the therapy of Graves' disease. J. Clin. Invest. **21** (1942), 624.
70. *Herzmann, H., A. Herzmann:* Zur Markierung von Humanserumalbumin mit Technetium-99. Isotopenpraxis 1 (1965), 130.
71. *Honda, T., I. Kazem, M. N. Croll, L. W. Brady:* Instant labelling of macro- and microaggregated albumin with ^{99m}Tc. J. Nucl. Med. **11** (1970), 580.
72. *Hosain, F., P. A. Intyre, K. Poulose* et al.: Binding of trace amounts of ionic indium-113m to plasma transferrin. Clin. Chim. Acta **24** (1969), 69.
73. *Hosain, F., P. Som, F. H. Deland, A. E. James, H. N. Wagner:* Yb-169-Calcium-DTPA: a new agent for cisternography. J. Nucl. Med. **11** (1970), 328.
74. *Huff, R. L., T. G. Hennersy, R. E. Austin, J. F. Garcia, B. M. Roberts, J. H. Lawrence:* Plasma and red cell iron tumors in normal subjects and in patients having various hematopoietic disorders. J. Clin. Invest. **29** (1950), 1041; **30** (1951), 1512.
75. *Jeghers, O., J. Abramovici, M. Jonckheer, A. M. Ermans:* A chemical method for the labeling of fibrinogen with ^{99m}Tc. Eur. J. Nucl. Med. **3** (1978), 95.
76. *Kaempfer, I., G. Schneider, A. Blottner, H. Deckart:* Markierung von Bilirubin mit Tc-99m und pharmakokinetische Untersuchungen. Radiobiol. Radiother. (Berlin) **23** (1982), 467.
77. *Kato-Azuma, M., M. Hazue:* Preparation and biologic evaluation of lipophilic derivatives of Tc-99m(Sn) Pyridoxylidenephenylalanine: an approach to structure/biodistribution relationship of technetium complexes. J. Lab. Comp. Radiopharm. (1981), 133.
78. *Kitani, K., G. V. Taplin:* Catabolic pathway differences between ^{131}I and ^{99m}Tc labeled albumin colloids and microaggregates. J. Nucl. Med. **13** (1972), 313.
79. *Klingensmith, W. C., A. R. Fritzberg, W. M. Spitzer* et al.: Clinical evaluation of Tc-99m-N,N,-bis (mercaptoacetyl)-2,3-diamino propionate (component A) (Tc-99m-CO_2-DADS-A) as a replacement für I-131-hippuran. J. Nucl. Med. **24** (1983), P 80 (abstr.).
80. *Knipping, H. W., W. Bolt, H. Venroth, H. Valentin, H. Ludes, P. Endler:* Eine neue Methode zur Prüfung der Herz- und Lungenfunktion. Die regionale Funktionsanalyse in der Lungen- und Herzklinik mit Hilfe des radioaktiven Edelgases Xenon-133. Dt. med. Wochenschr. **80** (1955), 1146.

81. *Kung, H. F., K. M. Tramposch, M. Blau:* A new brain perfusion imaging agent: ^{123}I-HIPDM: N,N,N'-Trimethyl-N'-(2-Hydroxy-3-Methyl-5-Jodbenzyl)-1,3 Propaanediamine. J. Nucl. Med. **24** (1983), 66.
82. *Lawrence, J. H., K. G. Scott, L. W. Tuttle:* Studies on leukemia with the aid of radioactive phosphorus. New Int. Clinics **3** (1939), 33.
83. *Lebowitz, E., M. W. Greene, R. Fairchild, P. Bradley-Moore, H. L. Atkins, A. N. Ansari, P. Richards, E. Belgrave:* Thallium-201 for medical use. J. Nucl. Med. **16** (1975), 151.
84. *Leeksma, C. H. W., J. A. Cohen:* Determination of the life span of human blood platelets using labelled diisopropylfluorophosphonate (DF^{32}P). J. Clin. Invest. **35** (1956), 964.
85. *Lin, T. H., A. Khentigan, H. S. Winchell:* ^{99m}Tc-Dimercaptosuccinic acid for renal imaging. J. Nucl. Med. **15** (1974), 512.
86. *Lin, T. H., A. Khentigan, H. S. Winchell:* A ^{99m}Tc labelled replacement for ^{131}I-rose bengal in liver and biliary tract studies. J. Nucl. Med. **15** (1974), 613.
87. *Loberg, M. D., M. Cooper, E. Harvey, P. Callery, W. Faith:* Development of new radiopharmaceuticals based on N-substitution of iminodiacetic acid. J. Nucl. Med. **17** (1976), 633.
88. *Loebe, J., I. Kaempfer, H. Deckart, J. Weiland:* Erfahrungen mit ^{113m}In-markierten Erythrozyten zur Milzszintigraphie. Radiobiol. Radiother. **16** (1975), 761.
89. *Malyshev, K. V., V. V. Smirnov:* A generator of gallium-68 based on circonium hydroxide. Radiokhimiya **17** (1957), 137.
90. *Maxfield, J. R., J. G. Maxfield, W. S. Maxfield:* Use of radioactive phosphorus and testosterone in metastatic bone lesions from breast and prostate. South. Med. J. **51** (1958), 320.
91. *McAfee, J. G., H. N. Wagner:* Visualization of renal parenchyma by scintiscanning with ^{203}Hg-neohydrin. Radiology **75** (1960), 820.
92. *McAfee, J. G., C. F. Fueger, H. S. Stern* et al.: ^{99m}Tc-pertechnetate for brain scanning. J. Nucl. Med. **5** (1964), 811.
93. *McAfee, J. G., H. S. Stern, G. J. Fueger, M. S. Baggish, G. B. Holzmann, I. Zolle:* ^{99m}Tc labeled serum albumin for scintillation scanning of the placenta. J. Nucl. Med. **5** (1964), 936.
94. *McFairlaine, A. S.:* Efficient trace-labelling of proteins with Iodine. Nature **182** (1958), 53.
95. *Metz, O., W. Stoll, W. Plenert:* Meningosis-Prophylaxe mit Radiogold (^{198}Au) bei der Leukämie im Kindesalter. Dt. med. Wochenschr. **102** (1977), 43.
96. *Mueller, J. H.:* Über die Verwendung von künstlichen radioaktiven Isotopen zur Erzielung von lokalisierten biologischen Strahlenwirkungen. Experientia **1** (1945), 199.
97. *Myers, W. G., J. C. Vanderleeden:* Radioiodine-125. J. Nucl. Med. **1** (1960), 149.
98. *Myers, W. G., H. O. Anger:* Radioiodine-123. J. Nucl. Med. **3** (1962), 183.
99. *Nishiyama, H., E. Deutsch, R. J. Adolph, V. J. Sodd, K. Libson, E. L. Saenger, M. C. Geson, M. Gabel, S. Lukes, J. L. Vanderheyden, D. L. Fortman, K. L. Scholz, L. W. Grossman, C. C. Williams:* Basal kinetic studies of Tc-99m-DMPE as a myocardial imaging agent in the dog. J. Nucl. Med. **23** (1982), 1093.
100. *Nouell, J. P., H. Renault, J. Robert et al.:* La bleomycine marquee au ^{57}Co. Nov. Presse Med. **1** (1972), 95.
101. *Nunez, J., J. Pommier:* New method of iodination and titration of proteins. Zürich. 10th Symp. on advances in tracer methodology, 1965.
102. *Olsson, C. G., U. Albrechtsson, L. Darte, R. B. R. Persson:* Tc-99m plasmin

for rapid detection of deep vein thrombosis. Nucl. Med. Commun. **3** (1983), 41.
103. *Pasqualini, R., G. Plassio, S. Sosi:* The preparation of albumin microspheres. J. Biol. Nucl. Med. **13** (1969), 80.
104. *Perez, R., Y. Cohen, R. Henry, C. Panneciere:* A new radiopharmaceutical for ^{99m}Tc bone scanning. J. Nucl. Med. **13** (1972), 788.
105. *Persson, R. B. R., V. Kempi:* Labelling and testing of ^{99m}Tc-streptokinase for the diagnosis of deep vein thrombosis. J. Nucl. Med. **16** (1975), 474.
106. *Poe, N. D., G. D. Robinson, L. S. Graham, N. S. MacDonald:* Experimental basis for myocardial imaging with ^{123}I-labeled hexadecanoic acid. J. Nucl. Med. **17** (1976), 1077.
107. *Pollähne, W., H. Deckart, J. Roemer:* ^{99m}Tc-Gelatine. Ein Radiopharmakon für die Leberszintigraphie. Radiobiol. Radiother. (Berlin) **11** (1970), 541.
108. *Pollycove, M., R. Mortimer:* The quantitative determination of iron kinetics and hemoglobin synthesis in human subjects. J. Clin. Invest. **40** (1961), 753.
109. *Potchen, E. J., M. Adatepe, M. Welch* et al.: Indium-113m for visualizing body organs. JAMA **205** (1968), 208.
110. *Pressman, D.:* Radiolabeled antibodies. Ann. N. Y. Acad. Sci. **69** (1957), 644.
111. *Rhodes, B. A., K. S. Turaihi, W. R. Bell, H. N. Wagner:* Radioactive urokinase for blood clot scanning. J. Nucl. Med. **13** (1972), 646.
112. *Robinson, G. D., A. W. Lee:* Radioiodinated fatty acids for heart imaging. Iodine monochloride addition compared with iodide replacement labeling. J. Nucl. Med. **16** (1975), 17.
113. *Rosa, U., G. Scassalati, F. Pennisi, G. Federighi, L. Donato, R. Bianchi:* Protein radioiodination by an eleytrolytic technique. Strahlentherapie **60** (1965), 258.
114. *Sarkar, S. D., W. H. Beierwaltes, R. D. Ice, G. P. Basmadijan, K. R. Hetzel, W. P. Kennedy, M. M. Masou:* A new and superior adrenal scanning agent NP. 59. J. Nucl. Med. **16** (1975), 1038.
115. *Schmidt, K. J., J. P. Hass, K. H. Brod, R. Wolf, H. Dietz:* ^{99m}Tc-Fe(II)-Komplex: eine neue Substanz zur Hirntumorszintigraphie. Atompraxis **14** (1968), direct inform 5.
116. *Schuemichen, C., K. Koch, A. Kraus, G. Kuhlicke, W. Weiler, A. Wenn, C. Hoffmann:* Binding of Technetium-99m to plasma proteine: influence on the distribution of Tc-99m phosphate agents. J. Nucl. Med. **21** (1980), 1080.
117. *Serafini, A. N., H. B. Hupf, D. Lindberg, W. M. Smoak, A. J. Gilson:* Iodine-123 rose bengal in the evaluation of the jaundiced patient. J. Nucl. Med. **16** (1975), 567.
118. *Sheppard, C. W., G. Jordan, P. P. Hahn:* Disappearance of isotopically labeled gold colloid from the circulation of the dog. Amer. J. Physiol. **164** (1951), 345.
119. *Sherman, A. J., M. Ter Pogossian:* Lymph node concentrative colloidal gold following interstitial injection. Cancer **6** (1953), 1238.
120. *Siegel, M. E., L. S. Malmud, B. A. Rhodes, W. S. Bell, H. N. Wagner:* Scanning of thrombemboly with ^{131}I-streptokinase. Radiology **103** (1972), 695.
121. *Sodee, D. B.:* A new scanning isotope: mercury-197. J. Nucl. Med. **4** (1963), 335.
122. *Spar, I. L., R. L. Goodland, S. J. Schwartz:* Detection of performed venous thrombi in dogs by means of ^{131}I-labelled antibodies to dog fibrinogen. Circulat. Res. **17** (1965), 322.
123. *Spicer, J. A., D. F. Preston, R. J. Baranczuk, E. Harvey, M. M. Guffey, D. L. Bradshaw, R. G. Robinson:* The biodistribution of Br-77 labeled Estrone: Visualization of tissues containing increased estrogen receptors. J. Nucl. Med. **20** (1979), 761.

124. *Stacy, B. D., G. D. Thorburn:* Chromium-51 ethylenediaminetetraacetate for estimation of glomerular filtration rate. Science **152** (1966), 1076.
125. *Stang, L. G. jr., W. D. Tucker, R. F. Doering, A. J. Weiss, M. W. Greene, O. H. Banks:* Development of methods for the production of certain short-lived isotopes. In: R. C. Extermann (Ed.): Radioisotopes in scientific research. Proc. Int. UNESCO-Conf., Paris 1957, 1, p. 50. Pergamon Press, New York 1958.
126. *Stern, H. S., D. A. Goodwin, H. N. Wagner, H. H. Kramer:* In ^{113m}—a short-lived isotope for lung scanning. Nucleonics **24** (1966), 57.
127. *Stern, H. S., D. A. Goodwin, N. Scheffel:* ^{113m}In for blood pool and brain scanning. Nucleonics **25** (1967), 62.
128. *Subramanian, G., J. G. McAfee:* A radioisotope generator of Indium-113m. Int. J. appl. Radiat. Isot. **18** (1967), 215.
129. *Subramanian, G., J. G. McAfee:* A new complex of ^{99m}Tc for skeletal imaging. Radiology **99** (1971), 192.
130. *Subramanian, G., J. G. McAfee, R. J. Blair, A. Mehter, T. Connor:* ^{99m}Tc-EHDP a potential radiopharmaceutical for skeletal imaging. J. Nucl. Med. **13** (1972a), 947.
131. *Subramanian, G., J. G. McAfee, A. Mehter, G. J. Blair, F. D. Thomas:* ^{99m}Tc-stannous phytate: a new in vivo colloid for imaging the reticuloendothelial system. J. Nucl. Med. **14** (1973), 459.
132. *Subramanian, G., J. G. McAfee, R. Blair, F. A. Kallfelz, F. D. Thomas:* Technetium-99m-methylene diphosphonate—a superior agent for skeletal imaging: Comparison with other Technetium-complexes. J. Nucl. Med. **16** (1975), 744.
133. *Taplin, G. V., O. M. Meredith, H. Kade:* The radioactive ^{131}I-tagged rose bengal uptake excretion test for liver function using gamma ray scintillation counting techniques. J. Lat. Clin. Med. **45** (1955), 665.
134. *Taplin, G. V., D. E. Johnson, E. K. Dorf, H. S. Kaplan:* Suspensions of radio-albumin aggregates for photoscanning the liver, lung, and other organs. J. Nucl. Med. **5** (1964), 259.
135. *Tauxe, W. N., M. K. Burbante, F. I. Maher, J. C. Hunt:* Renal clearances of radioactive ortho-iodo-hippurate and diatrizoate. Proc. Mayo Clin. **39** (1964), 761.
136. *Taylor, D. M., M. F. Cottrall:* The evaluation of amino acids labelled with ^{18}F for pancreas scanning. In: Radiopharmaceuticals and Labelled Compounds. Vol. 1, Vienna, IAEA, 1974, p. 433.
137. *Tewson, T. J., M. J. Welch, M. E. Raichle:* ^{18}F-labeled, 3-deoxy-3-fluoro-D-glucose: synthesis and preliminary biodistribution data. J. Nucl. Med. **19** (1978), 1339.
138. *Thakur, M. L., M. V. Merrich, S. W. Gunasekera:* Some pharmacological aspects of a new radiopharmaceutical, indium-111-bleomycin. In: Radiopharmaceuticals and labelled compounds. IAEA, Vienna 1973, p. 184.
139. *Thakur, M. L., M. J. Welch, J. H. Joist, R. E. Coleman:* Indium-111 labeled platelets: Studies on preparation and evaluation of in vitro and in vivo functions. Thromb. Res. **9** (1976), 345.
140. *Thakur, M. L.:* Gallium-67 and In-111 radiopharmaceuticals. Int. J. appl. Radiat. Isot. **28** (1977), 183.
141. *Thakur, M. L., J. P. Lavender, R. N. Arnot, D. J. Silvester, A. Segal:* Indium-111 labelled autologous leucocytes in man. J. Nucl. Med. **18** (1977), 1012.
142. *Thorell, J. I., B. G. Johansson:* Enzymatic iodination of polypeptides with ^{125}I to high specific activity. Biochem. Biophys. Acta **251** (1971), 363.
143. *Tjen, H. S. L. M., W. B. van der Pompe, P. H. Cox:* A new radiopharmaceutical for hepatobiliary imaging. Brit. J. Radiol. **50** (1977), 624.

144. *Tubis, M., E. Posnick, R. A. Nordyke:* Preparation and use of ^{131}J-labelled Iodium hippurate in kidney function tests. Proc. Soc. Exp. Biol. (N. Y.) **103** (1960), 497.
145. *Tubis, M., R. A. Nordyke, E. Posnick, W. H. Blahd:* The preparation and use auf ^{131}I-labelled sulphobromophthalein in liver function testing. J. Nucl. Med. **2** (1961), 282.
146. *Tubis, M., G. T. Krishnamurthy, J. S. Endow, W. H. Blahd:* ^{99m}Tc-penicillamine, a new cholescintigraphic agent. J. Nucl. Med. **13** (1972), 652.
147. *Verma, E. C., L. R. Bennett, J. J. Toya, D. L. Morton, E. Witt:* ^{111}In-Bleomycine: a new radiopharmaceutical for tumor scintigraphy. J. Nucl. Med. **14** (1973), 641.
148. *Volkert, W. A., D. E. Troutner, T. J. Hoffmann, R. M. Seger, R. A. Holmes:* Tc-99m-propyleneamine oxime (Tc-99m-PnAO); a potential brain radiopharmaceutical. J. Nucl. Med. **24** (1983), P 128.
149. *Wagner, H. N. jr., J. M. Weiner, J. G. McAfee, J. Martinez:* 1-mercuri-2-hydroxypropane (MAP). Arch. intern. Med. **113** (1964), 696.
150. *Wagner, S. J., M. J. Welch:* Gallium-68 labeling of albumin and albumin microspheres. J. Nucl. Med. **20** (1979), 428.
151. *Waxman, A. D., D. Tanacescu, J. K. Siemsen, R. S. Wolfstein:* Technetium-99m-glucoheptonate as a brain scanning agent. Critical comparison with pertechnetate. J. Nucl. Med. **17** (1976), 345.
152. *Wieland, D. M., J. I. Wu, L. E. Brown, T. J. Mangner, D. P. Swanson, W. H. Beierwaltes:* Radiolabeled adrenergic neuron-blocking agents. Adrenomedullary imaging with ^{131}I-Iodobenzylguanidine. J. Nucl. Med. **21** (1980), 349.
153. *Willvonsender, R. N., P. Goldstein, W. N. Tauxe:* Kinetik des radioaktiven Kupfers bei M. Wilson. Wiener Z. inn. Med. **54** (1973), 226 and 544.
154. *Winchell, H. S., W. D. Horst, L. Braun, W. H. Oldendorf, R. Hattner, H. Parker:* N-isopropyl-(^{123}I)p-iodoamphetamine: Single pass brain uptake and washout; binding to brainsynaptosomes; and localization in dog and monkey brain. J. Nucl. Med. **21** (1980), 947.
155. *Winkelmann, J. W., H. N. Wagner, J. McAfee, J. M. Mozley:* Visualization of the spleen in man by radioisotope scanning. Radiology **75** (1960), 465.
156. *Winsche, W. E., L. G. Stang, W. D. Tucker:* Production of ^{132}I. Nucleonics **8** (1951), 14.
157. *Winter, C. C., C. G. Myers:* Three new test agents for radioisotope renogram: DISA-I-BI, Cr-51-EDTA and Hippuran-I-125. J. Nucl. Med. **3** (1962), 273.
158. *Yano, Y., J. McRae, H. O. Anger:* Lung function studies using short-lived ^{81m}Kr and the scintillation camera. J. Nucl. Med. **11** (1970), 674.
159. *Yeh, S. H.:* ^{99m}Tc labeled toluidine-blue for liver scintigraphy. Int. J. appl. Radiat. **19** (1968), 885.
160. *Zolle, I., B. A. Rhodes, H. N. Wagner jr.:* Preparation of metabolizable radioactive human serum albumin microspheres for studies of the circulation. Int. J. appl. Radiat. Isot. **21** (1970), 155; Radiology **92** (1969), 1453.

Chapter 6

Radiopharmaceutical Agents: Recent Developments

Malcolm Frier

1. Introduction

In recent years, radiopharmaceutical agents have been produced with the object of demonstrating specific aspects of organ function. This has been accomplished by various modifications of physical or chemical characteristics, thus, for example, rendering the molecules highly lipid-soluble, or attaching the radioactive moiety to either particles of a highly specific size range, or to metabolically—active substrates. It is proposed to examine three main areas of development, although the lines of demarcation between the areas cannot be clearly drawn. Broadly, however, the areas are:

1. Radionuclides including the design and operation of new generator systems.
2. New formulations of technetium-99m.
3. A broad group of pharmacologically-active materials in general labelled with radionuclides other than technetium and including a discussion of the preparation and use of monoclonal antibodies.

Many of these developments have produced a clearer understanding of the structure/activity relationships of radiopharmaceutical agents, particularly of the Tc-99m-labelled compounds. Knowledge of the state of chemical combination of Tc-99m in compounds has broadened the scope of application beyond that of the highly polar materials whose distribution depends largely on passive diffusion across concentration gradients.

2. Generator systems

The best-established generator system to be introduced in recent years is the *Rubidium-81/krypton-81m* system. α-particle bombardment of a sodium bromide target produces a mixture of rubidium isotopes, predominantly rubidium-81, which decays to produce krypton-81m. The rubidium parent is adsorbed on to a column of zirconium phosphate, and the krypton daughter removed by passage of humidified air through the system. A method for improving the efficiency of use of krypton generators was recently introduced by Mostafa /23/ in which a reservoir breathing system was used (Figure 6.1.). The system is claimed by the authors to reduce the collection times by a factor of two when compared with three other delivery systems.

Although krypton-81m is now firmly established as a radionuclide for ventilation studies, it has been used as a tracer in perfusion studies of the lung, heart and brain. The main limitation of krypton in this respect results from its complete elimination by the lungs when injected intravenously.
Arterial angiography requires administration into the arterial supply of the organ perhaps requiring catheterization, which some would consider traumatic. Several other short-lived radionuclides which do not have this disadvantage are now under active consideration for perfusion studies including gold-195m and iridium-191m. Gold-195m generators are commercially available.
Gold-195m (half-life 30.5 s) is produced as the daughter of mercury-195m. The pair Hg-195m/Au-195m was explored as a potential generator system in 1974 but it is only recently that suitable elution systems have been devised to make the system viable. *Bett* /1/ proposed a system in which gold-195m was milked from a column of vicinal dithiol cellulose with potassium cyanide solution. Garcia /9/ proposed the use of a silica-gel column eluted with sodium thiosulphate and sodium nitrate. *Brihave* /2/ developed a system utilising a chelating resin eluted with 5 % dextrose in 0.01 M Trisma buffer at pH 7.7.

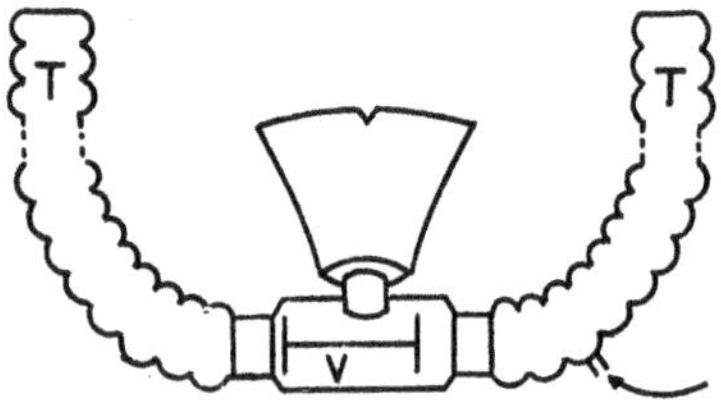

Fig. 6.1. Reservoir breathing system for delivery of Krypton-81m gas to a patient (Mostafa et al. 1983)
T = tubing of approx. 500 cc capacity V = 2-way non-return valve. Krypton gas inlet is arrowed

Iridium-191m is an ultra-short-lived radionuclide (half-life 4.9 s) that has found application in the evaluation of cardiac shunts in newborns /34/. Iridium-191m is the daughter product of osmium-191 which has a half-life of 15.4 days and is produced by the neutron irradiation of osmium-190. Early generators suffered from problems associated with poor yields, giving eluates with concentrations too low for adequate bolus injection. Osmium breakthrough was also higher than desirable. Recent developments have led to improved yields and to lower osmium breakthrough levels by the use of activated carbon columns and elution with 0.9 % sodium chloride solution containing 0.025 % potassium iodide. The 4.9-second half-life has been considered too short for the performance of satisfactory studies in adult patients. However, it has recently been reported by Packard /25/ that adequate studies are obtained if sufficient activity is injected.

3. Technetium-99m preparations

Developments in preparations containing technetium-99m have included the introduction of new classes of chemical compounds into clinical use, and the adoption of some established formulations for different clinical applications. Of the former group, interesting examples are the N,N'bis(S-benzoylmercapto-acetyl) ethylene diamines (DADS derivates) and the neutral, lipid-soluble agents having the property of crossing the blood/brain barrier. The latter group includes such materials as DTPA administered by aerosol for lung investigations, colloidal materials specifically for the investigation of bone marrow, and liposomes, which have been recently advocated as suitable for the detection of focal inflammatory lesions in deep-seated sepsis and rheumatoid arthritis.

3.1. DADS complexes

Following the development of this new class of ligand by Davison and Jones /4/ (1981) the complexes with technetium-99m were found to be rapidly excreted in the urine of rats. This property makes the materials potentially useful in the assessment of renal tubular function, in the same way in which iodine-labelled o-iodohippurate is used. The disadvantage they have is in the significant levels of activity excreted into the bile and subsequently found in the gastrointestinal tract.

In an attempt to improve the performance, Fritzberg /8/ synthesised several derivatives of DADS containing ethylene bridges, and found that the most promising of those studied was the carboxy analogue Tc-99m-N,N'-bis (mercapto-acetyl)-2,3-diaminopropanoate (CO_2-DADS-A). The structure of this compound is shown in Figure 6.2. The behaviour of this derivative was found to mimic that of "Hippuran" more closely than that of the DADS complex in experimental animals but in humans the urinary excretion was slower.

McAffee /18/ extended the range of derivatives studied by synthesising a further 21. Several of the compounds were evaluated in experimental animals in comparison with "Hippuran" and CO_2-DADS-A. In human volunteers, significant hepatic activity was still seen, and at one hour post-injection the plasma clearance and urinary excretion were much lower than that of simultaneously-administered o-iodohippurate. Klingensmith /11/ postu-

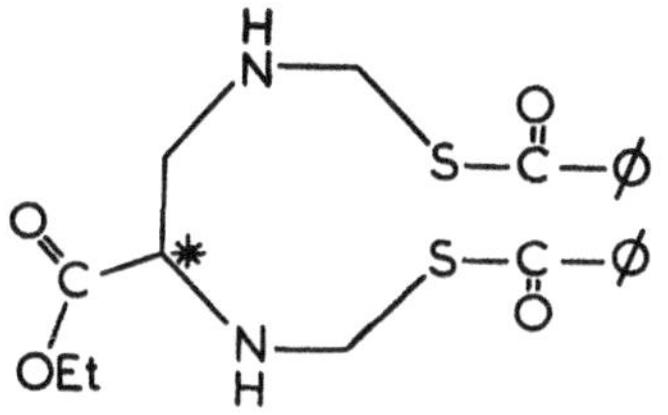

Fig. 6. 2. Chemical structure of N,N'-bis (mercaptoacetyl)-2,3-diaminopropanoate /CO_2DADS). The carbon atom at * is asymmetrical /18/.

lates that the slower plasma clearance and renal excretion of the DADS derivatives may be the result of a greater degree of protein binding and a smaller volume of distribution.

3.2. Lipid-soluble agents

A number of agents possessing high lipid solubility have been developed for the study of regional brain pathology. The first agent to be introduced for this purpose other than those containing positron-emitting radionuclides, was N-isopropyl-p-aminoamphetamine labelled with iodine-123 /38/, herein described. Subsequent developments have led to the study of lipophilic agents labelled with technetium-99m. Compounds such as aminoethanethiol, N,N'-diesters of EDTA and DTPA and bis-aminoethanol are highly lipophilic but show poor brain uptake, which may be related to high protein binding, or to poor in vivo stability. Two classes of compound have recently been extensively investigated as brain blood-flow agents; the derivatives of propylene amine oxime (PnAO) and of bis-aminoethanethiol (BAT). A series of BAT ligands was studied by Kung /15/. The proposed chemical structure for these compounds is shown in Figure 6.3. which also includes the substituents of the three derivates included in this study. Partition coefficients were high ranging from 56 to 541 and were the same when determined at pH 7.0 and 7.4 distinguishing them from the "pH shift" diamines labelled with iodine isotopes.

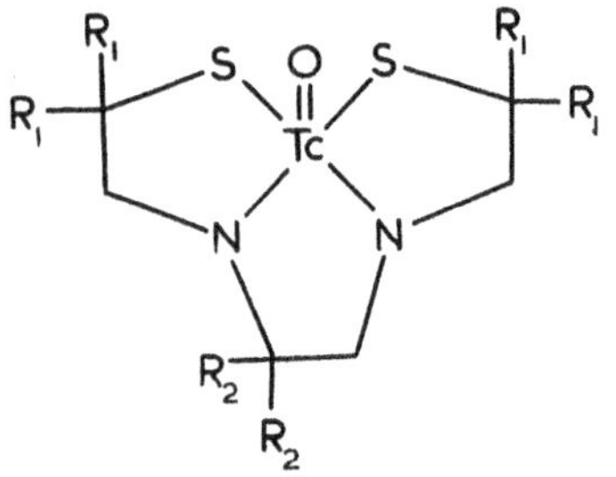

Fig. 6.3. Proposed chemical structure of 3 of the Tc-99m-1,2-dithia-5,8-diazacyclodecane (BAT) derivates studied by Kung /15/

	1	2	3
R1	CH	CH	C H
R2	H	CH	H

When studied in experimental animals, the distribution at 2 minutes following intravenous injection was similar to that expected from a knowledge of cardiac output. Brain uptakes for the three derivates and for iodine-labelled amines are shown in Table 6.1. Little activity is retained in the brain, indicating that the diffusion is a reversible process. High levels of protein binding (up to 83 %) have been demonstrated but this does not seem to affect brain uptake.
Derivatives of propylene amine oxine (PnAO) were suggested as potential brain radiopharmaceuticals by Volkert /35/. PnAO passively crosses the blood/brain barrier with blood-borne activity being rapidly cleared via the liver and kidneys. In animals, less than 8 % of the injected dose remains in

Table 6.1. Brain uptake in rats of 3 BAT derivatives, IMP and HIPDM ad 2 minutes post-injection. (Taken from the Data of Kung /15/.)

Radiopharmaceutical	BAT 1	BAT 2	BAT 3	IMP	HIPDM
Brain uptake (% injected dose)	1.96	2.77	2.92	2.64	2.74

the blood 2 minutes after injection. The most significant route of clearance is via the liver and bile. Analysis of the material appearing in the bile using HPLC shows identical peaks to those of the starting material, indicating a high in vivo stability. There is diffuse brain uptake within a few seconds of administration followed by rapid washout with a half-time of 40 s. Derivatives of this material, giving longer retention times in the brain are likely to form the basis of a range of technetium-99m labelled imaging agents.

3.3. Aerosol systems

Techniques for the evaluation of lung function using aerosol systems have been in regular use for a number of years. In 1970 Few and his co-workers /6/ described a spinning-disc system for generating solid aerosol particles of polystyrene at a controlled range of particle sizes. Such systems are of particular value in the study of mucociliary clearance.

Recent developments have concentrated on the preparation of solution aerosols for use as substitutes for radioactive gases in the study of lung ventilation. Of the available methods for the production of aerosols—jet nebuliser, ultrasonic, propellant, condensation, spinning disc and vibrating orifice—the simplest and most readily applied to lung ventilation is the jet nebuliser. A diagram of such a nebuliser is shown in Figure 6.4. The main disadvantage they have is in the range of droplet sizes produced. It is well established that particles larger than 5 μ in diameter are removed in the nasal cavity, while particles smaller than 0.1 μ remain suspended and are exhaled without deposition. An ideal droplet diameter is about 0.5 μ at which there is significant penetration of the peripheral airways followed by deposition. Royston /30/ describes a separator system constisting of a column packed with beads of various sizes, through which the aerosol must pass. Larger droplets impact on the beads and condense, thus allowing a suspension of droplets having a narrow size range to pass. Ventilation lung scans performed with aerosols generated using this separation device penetrate the small airways in a way comparable with that of krypton-81m gas.

The usual radiopharmaceutical agent chosen for nebulisation is technetium-99m DTPA. This is cleared from the lungs with a half-time of about one hour and has the advantage that, once absorbed from the lungs, it is rapidly excreted via the kidneys. This rapidly removes it from the field of view, thereby avoiding increased background counts when performing a subsequent perfusion scan.

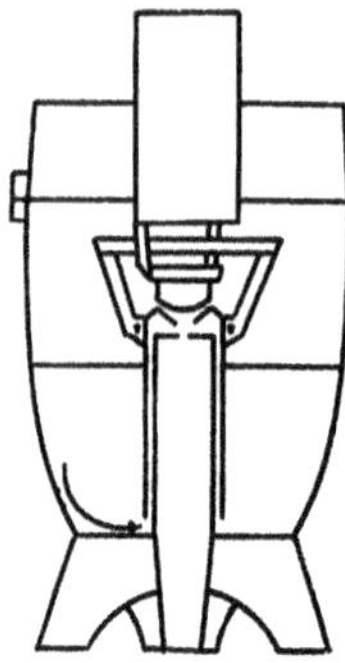

Fig. 6.4. Schematic diagram of one type of jet nebuliser. Liquid rises up the annular space (arrowed) surrounding the air-inlet tube, and is broken into droplets. Large droplets impact on the baffle. Small droplets are carried out with the airstream.

3.4. Small-particle colloids

The use of radioactive colloids for studies of the liver, spleen and lymph-nodes is well established. It is currently being advocated that colloids can also provide useful information at the very early stages of neoplastic involvement of the skeletal system, as bone-marrow scanning agents, and may in fact be superior to phosphonates in this respect. In a comparative study of bone-marrow and bone scanning Munz /24/ demonstrated that in 107 cases of patients with different malignant diseases, 95 (89 %) showed abnormalities on the bone-marrow scan whereas 72 (67 %) had abnormal bone scans.

Small-particle colloids of minimicroaggregated human serum albumin are now commercially available. In a comparative study of four colloids, McAfee /17/ placed them in rank order minimicroaggregated albumin > antimony sulphide > stannous phytate > sulphur colloid in terms of bone-marrow uptake. The actual amount in bone-marrow in a trial involving 15 patients was about 15 % of the administered dose. There is still substantial liver and spleen uptake of the order of 70—80 % of the injected dose and this can cause problems when dealing with marrow lesions in this area. However, even with this limitation, the technique is claimed to detect more malignant bone lesions than bone scintigraphy, and also to demonstrate their true extent and give earlier detection.

3.5. Liposomes

The use of liposomes as protective agents in drug delivery systems has been extensively investigated. In nuclear medicine, liposomes have been used as lymphoscintigraphic agents in the detection of metastatic spread of breast

cancer and have also been shown to detect experimentally-induced abscesses /21/. Richardson /29/ explored their use for the detection of tumours but the results were not encouraging.
A recent paper /22/ suggests the use of liposomes as a technique for the detection of deep-seated infection. Negatively-charged liposomes containing technetium-99m administered at a dose of 350 MBq showed uptake in a range of lesions including infected joints, infected hip prostheses, tubo-ovarian abscesses and joint affected by rheumatoid arthritis. There are significant limitations in the application of the technique to the detection of subphrenic or intrahepatic abscesses due to uptake by liver and spleen. These shortcomings are, however, shared by both gallium and indium-labelled leucocytes, and it would seem that liposomes merit further study in this application.

4. Metabolically-active materials

This group of substances is linked by the fact that they are not of the highly polar nature characteristic of the majority of Technetium-99m labelled agents in regular use. They are, in general, substances which if administered at appropriate concentrations would exert some pharmacological action, and enter into metabolic processes. The range of radionuclides used as labels includes iodine-131, iodine-123, selenium-75 and indium-111.

4.1. I-131m-iodobenzyl guanidine (MIBG)

In a study of radiolabelled adrenergic neuron-blocking agents, Wieland /36/ synthesised MIBG which is a guanethidine analogue capable of localising in the adrenal medulla. In structure it is similar to noradrenaline, and is taken up by the adrenergic storage vesibles. Clinical studies by Sisson /31/ and Sutton /32/ demonstrated the potential of the material as a means of identifying and localising disseminated malignant phaeochromocytoma. This is a difficult condition to identify by other techniques including computed axial tomography, intravenous contrast, and selective venous sampling, and scanning using MIBG has been recommended as the initial investigation of choice where phaeochromocytoma is suspected on clinical or biochemical grounds. The material is also advocated as a therapeutic agent in this condition. There is an indication that MIBG is also taken up by carcinoid tumours /7/.

4.2. 6-β-methylselenomethyl-19-norcholest-5(10)-en-3β-ol-Se-75

Experiences with iodinated derivatives of cholesterol have demonstrated the value of adrenal scanning in conditions such as Conn's Syndrome and Cushing's Syndrome. However, for a variety of reasons,including the instability of the agent, requirement for long scanning times and substantial radiation doses to patients, there are significant limitations to their use. The selenium-labelled agent has severel advantages over the iodine-labelled agent which have been summarized by Montz /20/ as follows:

a) Better counting efficiency (by a factor of 1.4) for selenium-75 vs. iodine-131.
b) Higher uptake by normal adrenal cortex and by hyperfunctioning glands or adenomas.
c) An enhanced difference (by a factor of 2) between radioactivity in adrenals and background activity.
d) Excellent stability of the agent-unaltered results have been obtained after storage at 4 degrees Celsius for more than 3 months.

When imaging patients in whom the presence of Cushing's Syndrome is suspeced, 7.4 MBq of the agent are administered on day 1. Posterior images are taken on days 3 and 6. It may be useful to localise the kidneys by administering a dose of 37 MBq technetium-99m DTPA. When investigating patients presenting with the symptoms of Conn's Syndrome, adrenal activity is supressed by administration of a course of Dexamethasone tablets 4 × 2 mg daily beginning 3 days prior to the administration of the selenium agent. Counting the start of the tablet course as day 1, imaging is performed on days 4 and 7. At this point the tablets are stopped and a further image taken on day 14. Localization of the kidneys may again be done with technetium-99m DTPA. This same agent may also be used as a subtraction agent to improve the adrenal images, and examples of images produced in this way are shown in Figure 6.5.

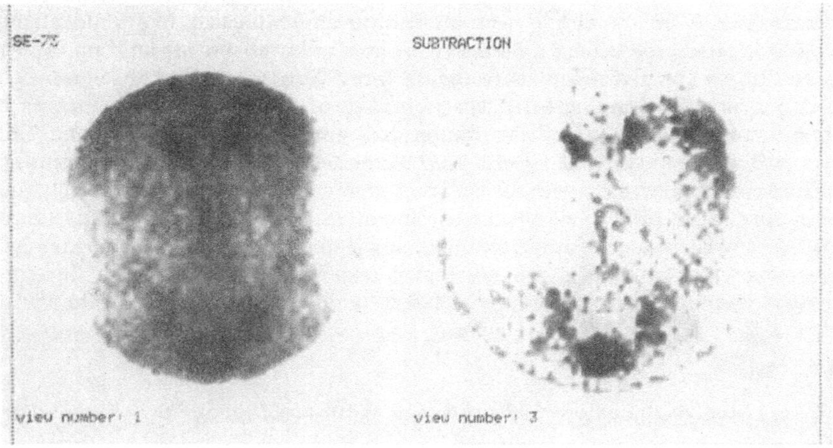

Fig. 6.5. Adrenal glands in a 59-year-old male patient suspected of having Conn's Syndrome. View 1 was taken on day 14. View 3 is the same view following subtraction with Tc-99m DTPA. No abnormality was detected.

4.3. N-isopropyl-p-(I-123)-iodoamphetamine (IMP) and N,N,N'-trimethyl-N'-(2-hydroxy-3-methyl-5-(I-123)-iodobenzyl-1,3-propane diamine (HIPDM)

The monoamine IMP was the first agent to be introduced for single-photon-emission tomography in brain blood-flow studies. Its mechanism of localisation in the brain has been proposed as being due to the presence of high capacity, non-specific amine-binding sites /38/. Evidence for this is that N-isopropyl-p-iodoamphetamine will inhibit the uptake of serotonin and norepinephrine in synaptosomes.

Diamines such as HIPDM change from an unionised to an ionised form with change of pH. When intracellular pH is significantly lower than blood pH, the agents, being weak bases, become charged. Hence their lipid solubility is reduced and they become trapped in the cell. The structures of IMP and HIPDM are shown in Figure 6.6. The radioiodine in each molecule is attached to a benzene ring, rendering the compounds stable to in vivo deiodination.

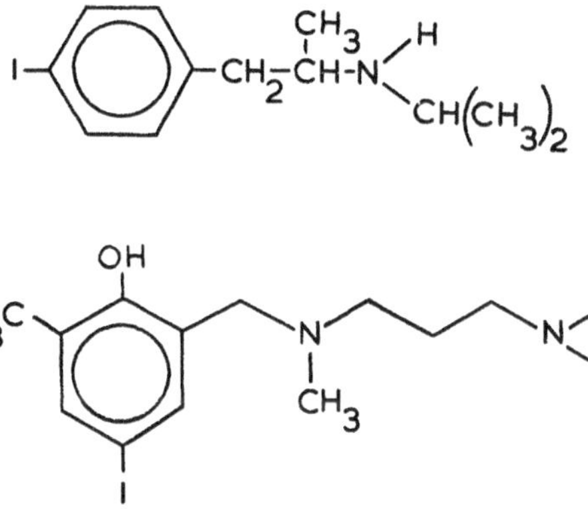

Fig. 6.6. Chemical structures of IMP and HIPDM. Note that in each case the iodine is attached to a benzene ring.

Following administration, the kinetics of IMP and HIPDM are very similar. There is an initial accumulation in the lung reducing to 50 % of maximum at 1 hour for HIPDM and 30 % of maximum for IMP. Brain uptake of IMP is higher than that of HIPDM, but the latter shows a substantially more rapid rate of accumulation in the brain. At 2 minutes after injection, some 75 % of the maximum brain activity is reached with HIPDM, whereas IMP reaches only 45 %.

The initial distributions of both IMP and HIPDM are directly proportional to regional blood flow. Quantification of blood flow is possible on the basis of certain assumptions which have been described by Kuhl /14/. The mathematical model described requires that:

a) the tracer must be delivered continuously to the brain.
b) it must be freely diffusible and be completely removed at a single pass.
c) its rate of back-diffusion from the brain to the blood must be slow.
d) it must be distinguishable from its polar metabolites.

The technique requires continuous arterial blood sampling over a period of 5 minutes post-injection. Simpler methods have been suggested. Hill /10/ put forward the idea that the absolute percent of injected dose of IMP as measured by emission tomography is proportional to regional cerebral blood flow. Severel more assumptions are necessary e.g. rate of clearance from the lungs must be constant between patients, and extraction of tracer by other organs should be constant, but the need for arterial blood sampling is eliminated.
Single-photon-emission tomography using the iodinated amines has proved a useful technique in a variety of conditions, and can give positive results at an earlier stage than X-ray CT methods in stroke patients. The method is also useful in assessing the results of surgery as, for example, following the treatment of carotid artery stenosis. Magistretti /16/ has examined the distribution of IMP in a group of patients suffering from epilepsy and has found good correlation between the site of abnormal focal increase of activity and the final location of the seizure focus.

4.4. Monoclonal antibodies

The concept of producing tumour-specific agents for either diagnostic or therapeutic applications has led to a great deal of research and development into the preparation and use of monoclonal antibodies. The specificity of antibodies has been known for many years. Antibody molecules in general vary little in their structure but their specificity is due to the amino acid sequence in a small region of the molecule known as the complementarity region. Their remarkable specificity would appear to render them ideal radiopharmaceutical agents. However, until comparatively recently, their availability was extremely limited because of the preparation techniques. They were available generally as complex protein mixtures isolated from animal sera. Standardisation was difficult, and there were potential immunogenic problems asociated with the protein mixture itself.
Many of these problems have been overcome by the introduction of monoclonal antibodies produced in cell culture, based on the hybridisation techniques described by Köhler and Milstein /12/. The usual technique adopted is to immunise mice with cells taken from tumours, following which the spleens are removed. The immune spleen cells are hybridised with mouse melanoma cells, resulting in the production of a prolific cell line capable of being grown in cell culture. By cloning, hybridoma cells are produced which are the daughters of single cells. The antibodies produced from these cells are therefore uniform in structure and in antigen-binding properties and are capable of being produced in virtually limitless quantities. By choosing the initial immune challenge presented to the host animal, it has been

possible to produce monoclonal antibodies specific for, among others, colorectal tumours, melanoma and osteogenic sarcoma.
Radiolabels have included iodine-131, iodine-123 and indium-111 and some work has been done with gallium-67 and technetium-99m. Most of the clinical work reported to date has used iodine-131 or indium-111, and labelling techniques for the introduction of these radionuclides will now be described.

Iodination
Standard iodination techniques such as the use of iodine monochloride, chloramine T or hypochlorite have been used to introduce iodine isotopes into antibody molecules. The major problems concern the effects of the rather harsh conditions required on the specificity of the labelled antibody produced. Good results have been achieved using the poorly-soluble oxidising agent iodogen (1,3,4,5-tetrachloro-3α6α-diphenylglucouril). This is usually coated as a solid phase on to plastic reaction tubes to which the mixture of antibody and iodide in a suitable buffer is added. Because oxidant is released into the reaction mixture only at the levels necessary for the reaction, there is no excess oxidising agent present to cause protein damage, and high levels of specificity are retained by the antibody. Recently Millar (1983) has described a technique where the iodogen is used as a solid suspension rather than a coating.

Indium labelling
The incorporation of cationic radionuclides into proteins requires that a conjugating agent be introduced which will act as a link between the two. Krejcarek and Tucker /13/ described the covalent attachment of chelating groups to macromolecules, and published methods for DTPA and EDTA. Perkins /27/ describes the labelling of the monoclonal antibody 791T/36 (raised against an osteogenic sarcoma) using the cyclic dianhydride of DTPA. Conjugation of the DTPA anhydride with the antibody takes place at pH 7 in phosphate-buffered saline, followed by removal of excess DTPA anhydride by gel filtration. Indium chloride (In-111) is added after dilution 1 : 1 with 1 M sodium acetate solution, followed by a further gel filtration stage to remove free indium. Reaction conditions are therefore mild with little or no loss of specificity. Some aggregation may take place during the labelling procedure.
There are significant advantages in using indium-111 as the radiolabel as opposed to iodine-131. Detection of indium using gamma cameras is more efficient because of its lower energy and the fact that thinner collimators can be used. Radiation dose to the patient is significantly lower. A large proportion of the injected dose of radioiodine-labelled preparation remains in circulation reducing the uptake ratio of tumour site to background, and necessitating the use of subtraction techniques. An example is shown in Figure 6.7. Free iodide is released, making it necessary to administer thyroid-blocking agents, and also further degrading the image quality by being taken up in stomach and urinary tract.
Owing to the multiplicity of interactions that can occur between tumour-cells and antibodies, specificity is not as high as might be expected from in vitro tissue culture studies. This means that antibodies raised against one

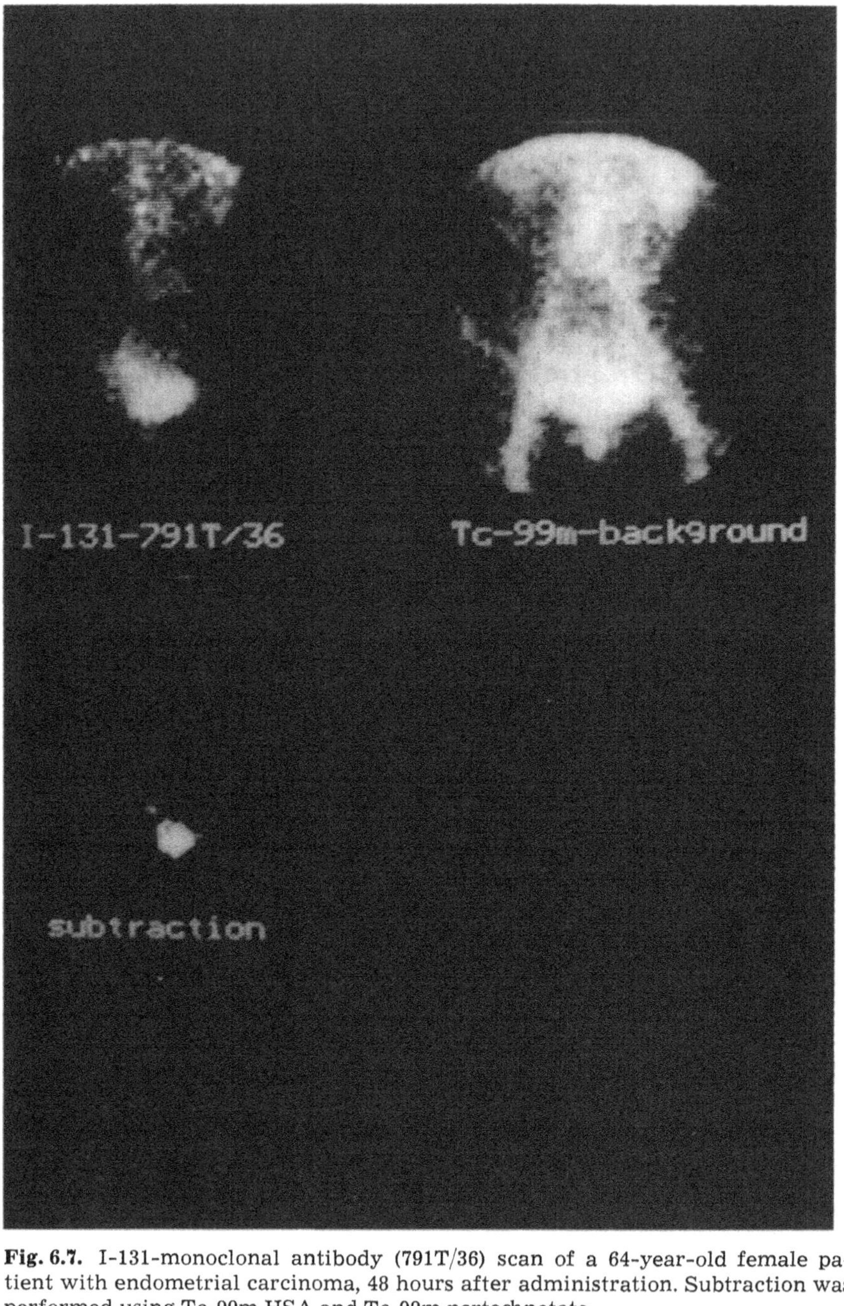

Fig. 6.7. I-131-monoclonal antibody (791T/36) scan of a 64-year-old female patient with endometrial carcinoma, 48 hours after administration. Subtraction was performed using Tc-99m HSA and Tc-99m pertechnetate.

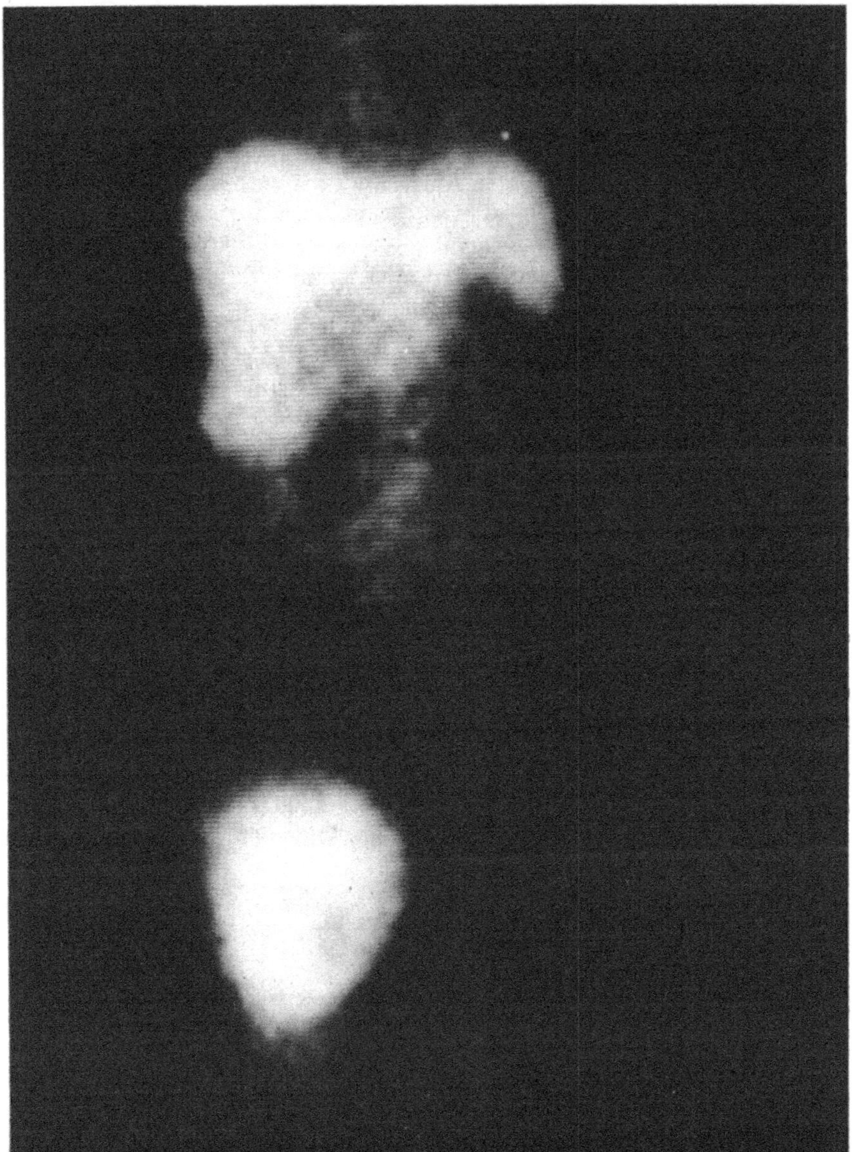

Fig. 6.8. In-111 monoclonal antibody (791T/36) scan of 70-year-old-male patient with a malignant fibrous histiocytoma of the right thigh, 72 hours after administration. Upper scan – anterior view of the abdomen showing liver uptake. Lower scan – anterior view of thighs showing intense uptake at the tumour site.

type of tumour can show significant uptake in other tumour types. The osteogenic sarcoma antibody 791T/36 has been studied in patients presenting with very different tumour types, e.g. bone /5, 26/, breast /34/ and ovary /33/. Figure 6.8. shows uptake of indium-111 labelled antibody in the right thigh of a 70-year-old male patient diagnosed as having a malignant fibrous histiocytoma. It is rather too early to suggests that imaging can provide the surgeon with valuable information concerning the extent of spread of malignant disease, and can be helpful in assessing the recurrence of previously-treated disease. They may also have a role to play in therapy, either in the targetting of chemotherapeutic agents or therapeutic doses of β-emitting radionuclides.

Antibodies of much larger molecular weight than 791T/36 have been evaluated clinically. 791T/36 has a molecular weight of 150,000 Daltons. C14/1/46/10, an antibody of the IgM class raised against a human colonic adenoma /3/ and C24, also an IgM class antibody raised against a human colorectal carcinoma /28/ both have molecular weights of 900,000 Daltons. In vitro, high levels of preferential uptake to membranes of human colorectal carcinomas can be demonstrated. Clinically, 9 patients studied with C14 and 5 patients studied with C24 have given negative results. However, these materials do not extravasate, and it is thought that they cannot gain access to the appropriate receptor sites. A possible solution to this problem would be to produce fragments of the antibody molecules which would retain the specificity but be of a much lower molecular weight. Considerable experience has been gained in the production and use of fragments, and these will now be considered.

In order to describe the production of fragments, it is necessary to examine the structure of the antibody molecule. The basic unit or monomer of an immunoglobulin molecule comprises 4 polypeptide chains (Figure 6.9.), the H (heavy) chains and the L (light) chains. The so-called V (variable) regions of these chains are small groups of amino acids which define the specificity of

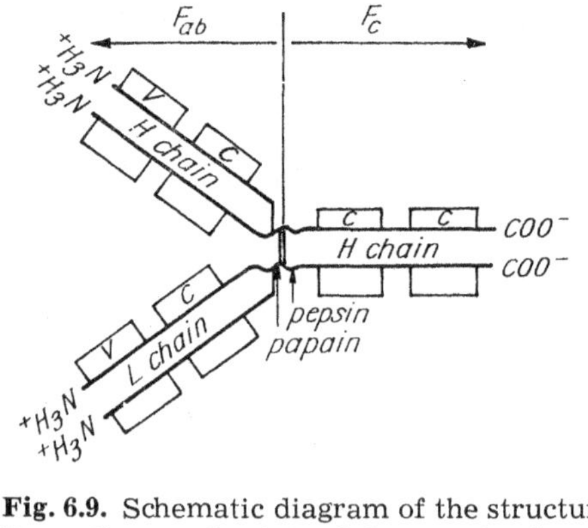

Fig. 6.9. Schematic diagram of the structure of an IgG molecule showing Fab and Fc portions and points of cleavage by the enzyme pepsin and papain. C = constant regions, V = variable regions.

the antibody and bind antigen. Digestion of an IgG molecule by the enzyme papain produces two Fab (antigen binding) fragments and one Fc fragment. These Fab fragments are termed monovalent in that they only possess one binding site. Digestion with pepsin produces one F (ab')^2 molecule together with small peptides. This fragment is termed divalent and possesses two binding sites.

Clinical experience with radiolabelled monovalent fragments has shown them to be cleared rapidly from the bloodstream and to give high renal uptake. Uptake into tumour sites is low. Results with divalent fragments have been more encouraging, presumably because both binding sites are retained. Tumour uptake is therefore higher, while still retaining the advantage of rapid blood clearance.

References

1. *Bett, R., G. H. Colman, J. G. Cunninghame:* Preparation of a Hg-195m/Au-195m isotope generator for medical use. Nucl. Med. Commun. **2** (1981), 75.
2. *Brihaye, C., M. Guillame, N. Lavi, M. Cogneau:* Development of a reliable Hg-195m/Au-195m generator for the production of Au-195m, a short-lived radionuclide for vascular imaging. J. Nucl. Med. **23** (1982), 1114.
3. *Brown, A., T. Feizi, H. C. Gooi, M. J. Embleton, J. K. Pickard, R. W. Baldwin:* A monoclonal antibody against human colonic adenoma recognises defucosylated Type-2-blood chains. Bioscience Reports **3** (1983), 163.
4. *Davison, A., A. G. Jones, C. Orvig:* A new class of oxotechnetium (5+) chelate complexes containing a $TcON_2S_2$ core. Inorg. Chem. **20** (1981), 1629.
5. *Farrands, P. A., A. C. Perkins, M. V. Pimm, J. G. Hardy, M. J. Embleton, R. W. Baldwin, J. D. Hardcastle:* Radioimmunodetection of human colorectal cancers by an anti-tumour monoclonal antibody. Lancet II (1982), 397.
6. *Few, J. D., M. D. Short, M. L. Thompson:* Preparation of Tc-99m-labelled particles for aerosol studies. Radiochem. Radioanal. letters **5** (1970), 275.
7. *Fischer, M., D. Kamanabroo, H. Sonderkamp, T. Proske:* Scintigraphic imaging of carcinoid tumours with I-131-metaiodobenzylguanidine. Lancet II (1984), 165.
8. *Fritzberg, A. R., C. C. Kuni, W. C. Klingensmith:* Synthesis and evaluation of a Tc-99m-N,N'-bis(mercaptoacetyl)-2,3-diamino-propionate: a potential replacement for [I-131] o-iodohippurate. J. Nucl. Med. **23** (1982), 592.
9. *Garcia, E., I. Mena, R. DeJong:* Gold-195m. Short-lived single photon emitter for haemodynamic studies. J. Nucl. Med. **22** (1981), 71.
10. *Hill, T. C., B. L. Holmann, R. Lovett:* Initial experience with SPECT of the brain using N-isopropyl [I-123]p-iodoamphetamine. J. Nucl. Med. **23** (1982), 191.
11. *Klingensmith, W. C. III, A. R. Fritzberg, V. M. Spitzer, D. L. Johnson, C. C. Kuni, M. R. Williamson, G. Washer, R. Weill, III:* Clinical evaluation of Tc-99m-N,N'-bis(mercaptoacetyl)-2,3-diaminopropanoate as a replacement for I-131-hippurate. J. Nucl. Med. **25** (1984), 42.
12. *Köhler, G., C. Milstein:* Continuous culture of fused cells producing antibody of predefined specificity. Nature (London) **256** (1975), 495.
13. *Krejcarek, G. E., K. L. Tucker:* Covalent attachment of chelating groups to macromolecules. Biochem. Biophys. Res. Comm. **77** (1977), 581.
14. *Kuhl, D. E., J. R. Barrio, S. C. Huang, C. Selin, R. F. Ackermann, J. L. Lear, J. L. Wu, T. H. Lin, M. E. Phelps:* Quantifying local cerebral blood flow by

N-isopropyl-p-[I-123]-iodoamphetamine (IMP) tomography. J. Nucl. Med. **23** (1982), 196

15. *Kung, H. F., M. Molnar, J. Billings, R. Wicks, M. Blau:* Synthesis and biodistribution of neutral, lipid-soluble Tc-99m complexes that cross the blood/brain barrier. J. Nucl. Med. **25** (1984), 326.
16. *Magistretti, P., R. Uren, D. Shomer, H. Blume, B. Holmann, T. Hill:* Emission tomographic scans of cerebral blood flow using I-123-iodoamphetamine in epilepsy. In: Proceedings of the Third World Congress of Nuclear Medicine and Biology. Ed. C. *Raynaud*. Pergamon Press, Oxford 1982, p. 139.
17. *McAfee, J. G., G. Subramanian, T. Aburano, Thomas F. Deaver, P. Fernandez, G. Gagne, B. Lyons, C. Zapf-Longo:* A new formulation of Tc-99m-minimicroaggregated albumin for bonemarrow imaging: comparison with other colloids, In-111 and Fe-59. J. Nucl. Med. **23** (1982), 21.
18. *McAfee, J. G., G. Subramanian, R. F. Schneider, M. Roskopf, B. Lyons, C. Ritter, C. Zapf-Longo, E. Palladino, F. D. Thomas:* Technetium-99m DADS complexes as renal function and imaging agents. J. Nucl. Med. **26** (1985), 375.
19. *Millar, W. T., J. F. Smith:* Protein iodination using iodogen. Int. J. Appl. Rad. Isot. **34** (1983), 639.
20. *Montz, R., J. Hagemann, W. Mischke:* Clinical evaluation of a new Se-75-labelled agent for adrenal scintigraphy. Brit. J. Radiol. **51** (1978), 641.
21. *Morgan, J. R., K. E. Williams, R. L. I. Davies, M. Thompson, L. A. Williams:* Localisation of experimental staphylococcal abscesses by Technetium-99m labelled liposomes. J. Medical Microbiol. **14** (1981), 213.
22. *Morgan, J. R., L. A. Williams, C. B. Howard:* Technetium-labelled liposome imaging for deep-seated infection. Brit. J. Radiol. **58** (1985), 35.
23. *Mostafa, A. B., P. O. Childs, D. A. Causer:* A quantitative evaluation of breathing systems used with Kr-81m generators. J. Nucl. Med. **24** (1983), 157.
24. *Munz, D. L., R. Kötter, I. Kornemann:* Bone marrow scanning in the early diagnosis of neoplastic involvement of the skeletal system: a comparative parallel study. Proc. 21st annual meeting of the Society of Nuclear Medicine (Europe), Ulm (Germany) 1983.
25. *Packard, A. B., G. M. O'brien, S. T. Treves:* The current status of the Os-191/Ir-191m generator. Proc. 2nd. European Symposium on radiopharmacy and radiaopharmaceuticals. Cambridge 1985.
26. *Perkins, A. C., N. C. Armitage, J. G. Hardy, M. L. Wastie, M. V. Pimm, J. D. Hardcastle:* Imaging bone disease using a radiolabelled anti-osteosarcoma monoclonal antibody. Nucl. Med. Commun. **5** (1984), 234.
27. *Perkins, A. C., M. V. Pimm, M. K. Birch:* The preparation and characterization of In-111-labelled 791T/36 monoclonal antibody for tumour immunoscintigraphy. Eur. J. Nucl. Med. **10** (1985), 1.
28. *Price, M. R., A. Brown, N. C. Armitage, R. W. Baldwin:* Application of a microradioimmunoassay to the analysis of monoclonal-antibody-defined epitopes on antigen preparations from human colonic cancer. ICRS Med. Sci. **13** (1985), 366.
29. *Richardson, V. J., B. E. Ryman, R. F. Jewkes, K. Jeysingh, M. N. H. Tattershall, E. S. Newlands, S. B. Kaye:* Tissue distribution and tumour localisation of Tc-99m-labelled liposomes in cancer patients. Brit. J. Cancer **40** (1979), 35.
30. *Royston, D., B. D. Minty, A. Houston, J. G. Jones, M. McCleod:* A simple separator to generate half micron particles for lung imaging. Brit. J. Radiol. **57** (1984), 223.
31. *Sisson, J. C., M. S. Frager, T. W. Valk:* Scintigraphic localisation of phaeochromocytoma. N. Engl. J. Med. **305** (1981), 12.
32. *Sutton, H., P. Wyeth, A. P. Allen, D. A. Thurtle, T. K. Haines, M. I. D. Cawley,*

D. Ackery: Disseminated malignant phaeochromocytoma: Localisation with iodine-131-labelled metaiodobenzylguanidine. Brit. Med. J. **235** (1982), 1153.

33. *Symonds, E. M., A. C. Perkins, M. V. Pimm, R. W. Baldwin, J. G. Hardy, D. A. Williams:* Clinical implications for immunoscintigraphy in patients with ovarian malignancy: a preliminary study using monoclonal antibody 791T/36. Brit. J. Obstet. Gynaecol. **92** (1985), 270.
34. *Treves, S., C. Cheng, A. Samuel:* First pass radionuclide angiography with ultra-short-lived Iridium-191m. J. Nucl. Med. **22** (1981), 8.
35. *Volkert, W. A., D. E. Troutner, T. J. Hoffmann, R. M. Seger, R. A. Holmes:* Tc-99m propyleneamine oxime (Tc-99m-PnAO): A potential brain radiopharmaceutical. J. Nucl. Med. **24** (1983), 128.
36. *Wieland, D. M., J. I. Wu., L. E. Brown, T. J. Mangner, D. P. Swanson, W. H. Beierwaltes:* Radiolabelled adrenergic neuron-blocking agents: adrenomedullary imaging with I-131-iodobenzylguanidine. J. Nucl. Med. **21** (1980), 349.
37. *Williams, M. R., A. C. Perkins, F. C. Campbell, M. V. Pimm, J. G. Hardy, M. L. Wastie, R. W. Blamey, R. W. Baldwin:* The use of Monoclonal Antibody 791 T/36 in the immunoscintigraphy of primary and metastatic carcinoma of the breast. Clin. Oncol. **10** (1984), 375.
38. *Winchell, H. S., W. D. Horst, L. Braun, W. H. Oldendorf, R. Hattner, H. Parker:* N-isopropyl-[I-123]-p-iodoamphetamine: single-pass brain uptake and washout: binding to brain synaptosomes and localisation in dog and monkey brain. J. Nucl. Med. **21** (1983), 947.

Chapter 7

Technetium-99m Radiopharmaceuticals: Structure-Activity Relationships

Rudolf Muenze

1. Introduction

At present ^{99m}Technetium radiopharmaceuticals are dominant in clinical nuclear practice, which is due to the favourable nuclear decay characteristics of ^{99m}Tc. Many of them, such as the "^{99m}Tc labelled" proteins, red blood cells, or complexonates, are simple volume radiotracers, which indicate the space and the fluid flow in certain compartments or detect lesions or barriers between different distribution spaces.

Their biodistribution is determined by physico-chemical properties, such as size, charge, and lipophilicity. A further group of valuable ^{99m}Tc-radiopharmaceuticals localise specific tissues depending on their biological state. They are suitable for the detection of pathological alterations in these tissues.

Most of these compounds were found empirically but under the assumption that the technetium labelled compound retains the characteristic biodistribution of the unlabelled molecule. Sometimes this assumption seems to be true, for instance with osteotropic phosphorous compounds or the kidney affine thiolato complexes. But most of the "labelled" molecules exhibit a quite different biological behaviour compared to the native substances. In favourable cases these altered properties could be used for the imaging of other organs. The well known hida chelates of Loberg /1/, originally designed in order to obtain a myocardial agent by labelling an analogue of myotropic lidocain, proved to be suitable hepatobiliary agents. This unsatifactory situation has already been recognized since 1977 when Eckelman /2/ wrote:

"The first prerequisite to assure further development of ^{99m}Tc-radiopharmaceuticals is the resolution of the many uncertainties concerning the chemistry of ^{99m}Tc.

Further elucidation of structure, oxidation state, equilibrium constants and kinetics associated with technetium chelates, and we must add, their relationships to biodistribution and biokinetics is needed."

Since that time a growing flood of experimental studies and interesting approaches has been published.

2. Technetium coordination chemistry – general features

In recent years chemical studies with ^{99}Tc model compounds have attracted a growing interest. But the question as to whether the results are conclusive in order to propose chemical structures for the unknown species in nanomolar ^{99m}Tc-radiopharmaceutical solutions has not been settled so far.

Commonly model compounds are assigned to ^{99m}Tc-radiopharmaceuticals if both have identical values in concentration independent properties like the distribution in chromatographic systems or the electrophoretic mobility and in their bio-distribution. But, unfortunately, different types of technetium chelates seem to exist. For certain ligands, such as the hydroxy carbonic acids, polyols, and thiolato compounds, we observe quite different properties with different chemical species of technetium with the same ligand. The second group comprises technetium complexes with aminopolycarbonic acids and with phosphonates. Different chemical species with the same ligand exhibit only minor differences in the above-mentioned properties /3/. This is demonstrated in Table 7.1.

The Tc(V)-dimercaptosuccinate complexes show striking differences in their biological behaviour depending on the stereochemistry of the ligand. Only the meso-complex proves to be osteotropic. The well known kidney affinity of the radiopharmaceutical ^{99m}Tc-DMSA is probably a peculiarity of the Tc(IV)-chelate (Figure 7.1.).

On the other hand, many osteotropic complexes with different phosphates and phosphonates all show bone uptake of about 2 % dose/gram organ in the rat independent of the oxidation state, the presence of tin in the complex, the molecular size and other properties /3/.

Thus, we have to be aware that conclusions which are drawn from ^{99}Tc model experiments with regard to the species in the corresponding radiopharmaceuticals have to be evaluated very critically in every case.

Only few years ago virtually no reliable information existed on the chemical state, composition, oxidation state, structure and reactivity of technetium complexes which have radiopharmaceutically relevant ligands, not even for ^{99}Tc model compounds. Now as an result of systematic studies carried out

Table 7.1. Different properties of ^{99}Tc- and ^{99m}Tc-ligand systems

^{99}Tc-model compounds millimolar	^{99m}Tc-radiopharmaceuticals nanomolar
similar physico-chemical properties (chromatogr. electroph.) →	
similar biodistribution, biokinetics →	

sensitive Tc-ligand systems	**insensitive Tc-ligand systems**
– hydroxycarbonic acids	– large biopolymers
– polyols	– cells
– thiolato compounds	– complexonates (EDTA, DTPA, HIDA etc.)
	– phosphates and phosphonates
different reaction products have *different* physico-chemical and biological properties	different reaction products have *similar* physico-chemical and biological properties

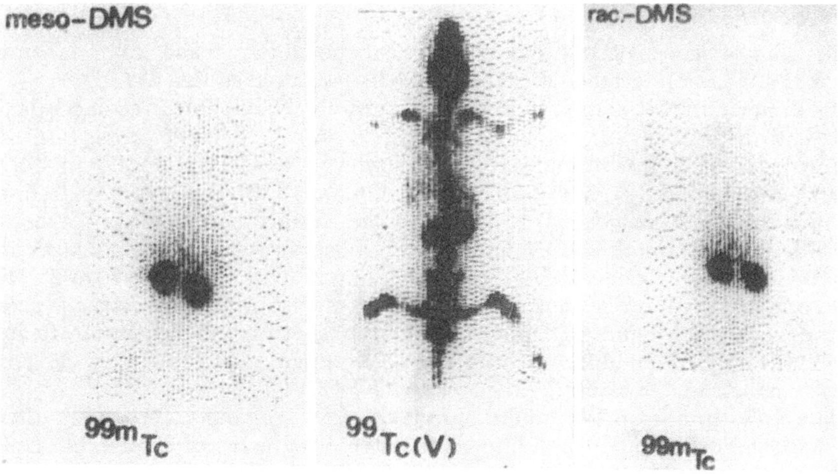

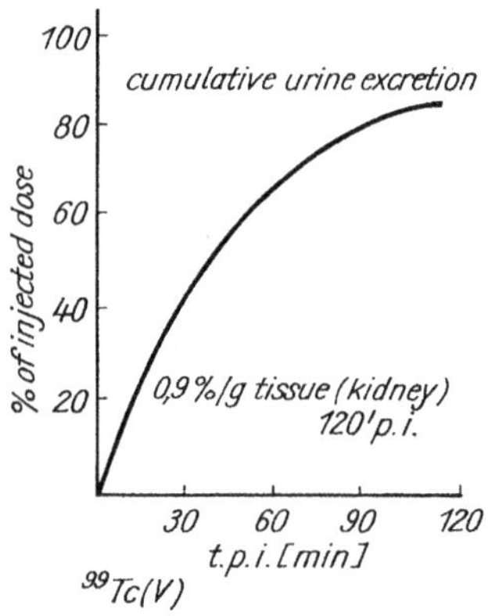

Fig. 7.1. Biological behaviour of dimercaptosuccinic acid (DMS)-complexes meso-DMS and rac-DMS

by several groups in different countries /5–23/ a considerable amount of detailed information is available and, therefore, we can try to summarize the most essential conditions for technetium complex formation in aqueous solution.

As to the oxidation states of technetium, certainly the most prominent forms in radiopharmaceutically relevant systems are Tc(V), Tc(IV) and Tc(III). This is in some respect unexpected because Tc(V) and Tc(III) have been considered to be very unstable oxidation states for a long time.

In most cases the ligand must possess at least two adjacent ligator groups which are stereo-chemically arranged in such a way that the formation of five or six membered chelate rings including the technetium atom are possible (Figure 7.2.).

Listed in Figure 7.2. are such coordination moieties for which the formation of "in vivo" stable technetium complexes in aqueous solution have been experimentally confirmed.

An additional condition for the stabilization of reduced technetium atoms in aqueous solution is a balanced ratio of charge- and orbital controlled metal interactions due to the technetium border-line character. Higher charged central atoms (Tc_5^+, Tc_4^+) prefer higher negatively-charged ligands such as oxo groups.

If neutral ligands are to be attached, then the technetium charge must be reduced by an appropriate number of oxo, hydroxo or other charged ligands. A distinguishable peculiarity of the oxidation state + 4 is its tendency to polymerize under formation of oxo or hydroxo bridges and frequent metal-metal bounds. The probability of this reaction is higher in ^{99m}Tc-carrier experiments than in the highly diluted ^{99m}Tc-radiopharmaceuticals.

Many of the isolated ^{99}Tc model compounds in this oxidation state proved to be polymeric, at least after prolonged standing. They are certainly not appropriate representatives for hypothetical species in radiopharmaceutical solutions. Soft neutral donors like arsines and phosphines prefer technetium in the + 3 or even + 2 state.

Following this short survey of more general chemical properties, relevant results will be summarized in order to derive some common properties for different classes of ligands.

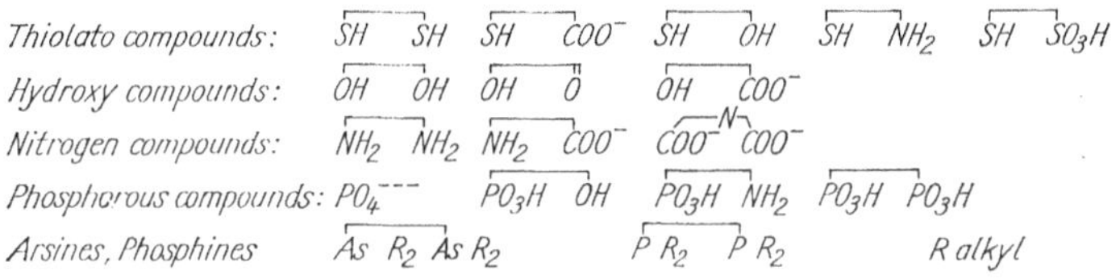

Fig. 7.2. Chelating moities for Tc complexing in aqueous solution

3. Technetium chelates with hydroxylic ligands

The first technetium coordination compounds which were introduced as radiopharmaceuticals—"Tc-citrate" and "Tc-Fe-ascorbinate"—were compounds with hydroxylic ligands /24, 25/.

The mechanism of reduction by tin(II) was the subject of many studies carried out by several authors, particularly by Steigman /26/ and others /27, 28/. Independent of the particular ligand, the reaction proceeds immeasurably quickly by formation of mononuclear bishydroxycarboxylato chelates. Dependent on the number of charged ligator atoms in the ligand, the necessary charge compensation is performed by the coordination of at least one oxo

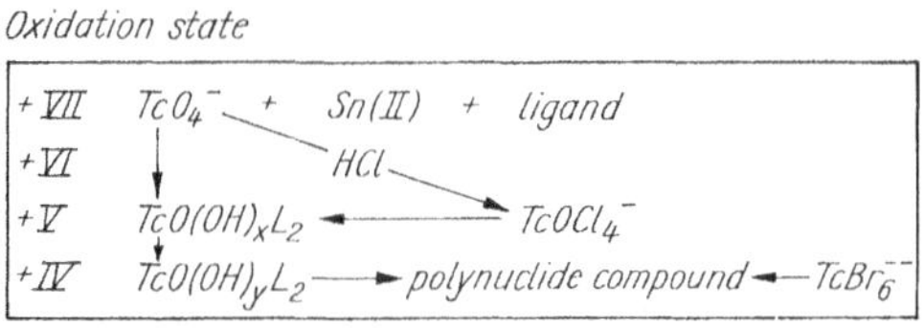

Fig. 7.3. Hydroxyl compounds-reaction scheme
H_2L = HO – Z – OH
HO – Z – COOH

and further oxo and/or hydroxo groups. The same compounds were produced by ligand exchange with $TcOCl_4^-$ (Figure 7.3.). The subsequent slow reduction of the Tc(V)-compound to the Tc(IV)-chelate leads to compounds of the same structure and charge, differing only by the protonation of one oxo group. As studied in the Tc-citrate system the Tc(V)/Tc(IV)-reduction is dependent on Tc(IV)-concentration /27/. It is an autocatalytic process and thus the reduction in the ^{99m}Tc solutions remains in the Tc(V)-state despite the huge excess of reductant in most cases but not in all. This fact is very important with respect to biological behaviour. It will be discussed in the second part of this chapter.
Table 7.2. demonstrates the variety of hydroxycarbonic and hydroxyl compounds studies by several authors.
The Tc(V)-complexes disproportion at higher concentration into TcO_4^- and the corresponding Tc(IV) chelate. For this reason, all attempts to isolate solid

Table 7.2. Hydroxyl compounds-Tc(V)-complexes chemically characterized in solution (charge, Tc/ligand ratio, size)

suggested structure
(ligand exchange to thiolato compounds)

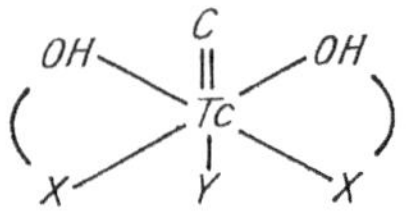

X	Ligand	Y
COO^-	α-hydroxiisobutanoic acid	O
	malic acid	OH
	tartaric acid	OH
	gluconic acid	OH
	glucoheptanoic acid	OH
	citric acid	COO^-
	isocitric acid	COO^-
OH	glycol	?
	glycerole	OH
	mannitol	?
	several carbohydrates and polyols including glucose, fructose, gelactose, desoxiribose, saccharose, starch	?

compounds have failed hitherto. The chemical and structural characterization in the diluted solutions was carried out using chemical and physicochemical methods such as the determination of reaction stoichiometries, genetic relationships in ligand exchange reactions, electrochemical determination of ionic charges, magnetic and spectrochemical measurements.
The Tc(V)-glucoheptonate and the Tc-citrate complexes were studied more in detail by Kieviets and also by our group /28, 29/. Pure hydroxyl compounds are stable only in the range of the deprotonation of at least one hydroxyl group, normally at alkaline pH /29, 30/. Only the higher polyols which are more acidic forms chelates at lower pH. With hydroxycarbonic acids mostly the neutral OH group together with the adjacent carboxyl group is coordinated, but with polyhydroxy carbonic acids, such as glucoheptanoic acid or gluconic acid coordination can also take place via ionized hydroxyl residues /28, 29/.
The Tc(IV)-chelates exist in monomeric forms at carrier concentration colourless, oligomeric pink or violet coloured and–highly polymeric–brown. The polynuclear compounds are produced spontaneously from the monomers. They are also directly obtained via ligand exchange with $TcBr_6^{--}$. The pink or violet substances have about two ligand atoms per technetium. A more detailed examination of the glassy violet Tc(IV)-citrate compound revealed its highly complicated structure as a trimeric oxygen-bridged metal cluster compound /31/. On the other hand, the brown types having about one ligand molecule per technetium atom are probably chain-linked oxygen-bridged polymers.
Reduction to the Tc(III)-state has been observed by polarographic methods both at the monomeric and polymeric Tc(IV)-hydroxycarbonic acid chelates /28, 32/. The oligomeric violet Tc(IV)-citrate forms a green Tc(III)-complex under rigorous Sn(II)-reduction /33/.

4. Technetium-thiolato complexes

The reduction scheme of pertechnetate in the presence of thiol compounds is quite similar to that of hydroxycarbonic chelates (Figure 7.5.). The reduction proceeds rather quickly with tin(II) and also with boranate and other reductants to the Tc(V)-state, and, moreover, also but retarded by the ligand itself. A particular feature in this group is the formation of the tris-aminobenzenthiol compound with the unusual trigonal prismatic structure which was observed by Spies /4/. In the presence of air this compound converts into a stable Tc(VI)-compound (Figure 7.4.).
The radiochemically relevant Tc(IV)-thiolato compounds have been completely uninvestigated so far.
The chemical structure of Tc(V)-thiolato compounds has been well-known since the exploration by X-ray analysis by Byrne et al. /34/. Quantum chemical calculations of this structure have shown that light absorption at about 400 nm of these yellow compounds is characteristic of the square technetium oxo structure /35/. That is why we assume that all the compounds

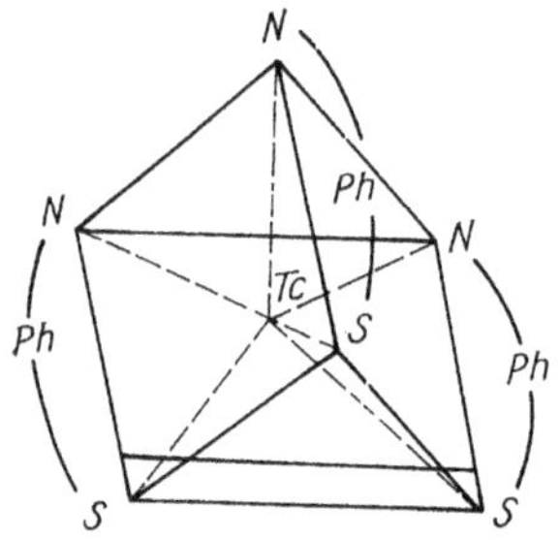

Fig. 7.4. Spherical figure of tris-aminobenzenethiolatotechnetium (VI)
Tc = technetium, Ph = phenyl group,
N = nitrogen, S = sulphur

listed in the table have the same principal molecular architecture. As demonstrated in Table 6 the second mercapto-group can be substitued by NH_2, OH, or COO^- (Table 7. 3., Figure 7.5.).

5. Technetium chelates with nitrogen compounds

The principal chemical properties of Tc-nitrogen chelates are similar to those of both groups mentioned above. They have attracted radiopharmaceutical interest recently by their cationic nature. The structure is probably similar to that of the corresponding Rhenium(V)-compounds which are distinguished by a linear 0 = Tc = 0 moiety surrounded by equatorially attatched nitrogen ligands. After the discovery of pure nitrogen chelates with cyclam by Troutner /36/, ethylenediamine by Thornback /37/ and pyridine by Kusina /12/, we found further Tc(V)-chelates /38, 39/ with more complicated pure nitrogen or mixed nitrogen-oxygen ligands, as it is shown in the next table. The complex formation takes place in the region where the deprotonated nitrogen donor exists. Optimal yields were obtained at the pK value of the nitrogen donor (Table 7.4.).

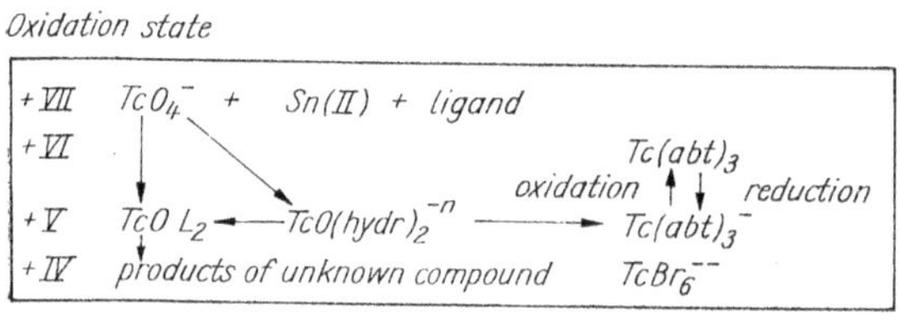

Fig. 7.5. Thiolato compounds – general reaction scheme
H_2L = HS – Z – SH
HS – Z – NH_2
HS – Z – OH
HS – Z – COOH
abt = aminobenzenethiol
hydr = hydroxycarbonic acid (glyconic, citric etc.)

Table 7.3. Thiolato compounds chemically characterzized in solution (charge, size, Tc/ligand ratio)

suggested structure
(electr. spectr.)

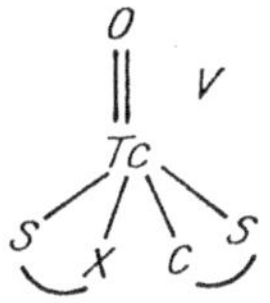

X	Ligand	
NH_2	cysteamine	cation
	cystein	anion
	cystein ethyl ether	anion
	penicillamine	anion
OH/COO^-	thioglycolic acid	anion
	thiolactic acid	anion
	thiomalic acid	anion
	3-mercaptopropionic acid	anion
	acetylcystein	anion
	gluthathion	
	mercaptoethanol	
	thioglucose	

Table 7.4. Nitrogen compounds – Tc(V)-complexes partially chemically characterized in solution

suggested structure
(analogy to $ReO_2\,en_2^+$, $ReO_2py_4^+$)

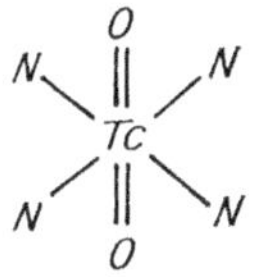

Ligand	Yield		Optimal pH	pK (nitrogen)$_{lig}$
Ethylenediamine	80–100	a)	10.5	7.1 / 10.7
	90– 95	b)		
β- bis -bimethylaminoethane	70–100	a)	10.2	(7 / 10)
Pyridine	90	a)	6.0	8.6
α, α'-dipyridine	80– 90	a)	5 – 7	9.7
	60– 70	b)	2.2– 7.5	
1,10-phenanthroline	50– 60	a)	4 – 6	4.8
	60– 65	b)	5 – 7	9.1
Aminopyridine	43	b)	5.5	6.85
Nicotinic acid amide	70	a)	4 – 5	
	60– 70	b)	6 –10	10.7
Thymine	37	b)	4	9.9 / 13
Cytosin	23	b)	2.1	4.6 / 12.2
Cyclam				

a) ligand exchange with $TcO_2hydroxyisobutanoic_2^+$
b) $^{99m}TcO_4^-$ reduction by Sn(II) in the presence of the nitrogen compounds

It should be emphasized that many drugs, aminoacids, nucleic acid compounds and other interesting biological substances can be labelled at the appropriate pH (Table 7.5.).
The diagnostic potential of this new group of technetium-chelates is completely unexplored.

6. Technetium chelates with aminopolycarbonic acids and phosphonates

The chemistry of Tc-aminopolycarbonic acid chelates and phosphonates is much more complicated and research into this field has just begun. The reduction of TcO_4^- in aminopolycarbonic acids seems to proceed via Tc(V) directly to Tc(III) in excess tin(II) (Figure 7.6.). But by stepwise addition of tin(II) Tc(IV) chelate is formed. A reasonable explanation of this fact is that the competition between the disproportionation of the intermediate Tc(V)-

Table 7.5. Nitrogen compounds – Tc(V)-complexes with unknown structure

Ligand	Yield		Optimal pH	pK (nitrogen)$_{lig}$
Sulfisomidine	76	b)	4.7	
Aminophenazone	27	b)	4.7	
Chinine	21	b)	7.5	5.9 / 9.9
Thiamine	27	a)	4.5	
	77	b)	4.5	
Ephedrine	17	b)	5	10.1
Urotropine	21	b)	4	
Lysine	45–85	a)	4.9	8.9 / 10.5
	40–65	b)	2.8	
Histidine	30–45	a)	5.4	8.2 / 12.2
	60–65	b)	2.4	
Ornithine	15–20	a)		
	18	b)	9–10	10.8 / 12.8

a) ligand exchange with $TCO_2hydroxyisobutanoic_2^+$
b) $^{99m}TcO_4^-$ reduction in the presence of the nitrogen compounds by Sn(II)

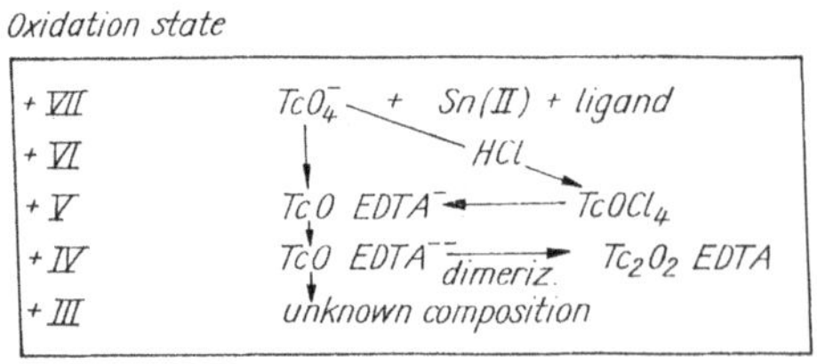

Fig. 7.6. Aminopolycarbonic acid compound – reaction scheme for polyaminopolycarbonic acids (EDTA, DTPA, etc.)

chelate into TcO_4^-- and Tc(IV)-chelate and Tc(V)/Tc(III)-reduction which depends on tin(II) concentration. But this has to be confirmed experimentally. The Tc(V)-state was observed transiently during the polarographic reduction of TcO_4^- in the solution of the ligand by Russell /40/. A Tc(V)-EDTA compound was synthesized by ligand exchange in a nonaqueous medium by Deutsch /41/. The isolated Tc-chelates with aminopolycarbonic acids from aqueous solutions are Tc(IV)-compounds /42/. They are produced either by stoichiometric reduction by tin(II) or by ligand exchange with $TcBr_6^{--}$. A more detailed investigation in our laboratory revealed a variety of reaction products /43, 44/. At first unstable uncoloured compounds were observed which converted into yellow monomeric and red and brown polymeric compounds respectively. With the terdentate iminoacetic acid and the quadridentate nitrilotriacetic ligands no monomeric products were found which is probably due to uncomplete shielding of the metal centre. In Table 7.6. we shall list the chemically characterized species, hitherto together with some confirmed or suggested molecular structures.

In ^{99m}Tc-radiopharmaceuticals technetium is probably in the Tc(III)-state. The chemistry of the radiopharmaceutically important chelates with phosphorous compounds is similar obscure. As in the group mentioned above reduction to Tc(V)-, Tc(IV)-, and Tc(III)-compounds has been observed /45, 46/. The only isolated compound of this group, an Tc(IV)-MDP complex, is a polymere bridged by the ligand /47/. Deutsch has shown /48/ that in model compounds produced by tin reduction this element was incorporated into the polymer structure. It is possible that the radiopharmaceutical compound is a polymer tin phosphonate with some technetium atoms in the chain. On the other hand, it is also possible that the radiopharmaceutical compound is a Tc(V) phosphonate. Recently Johannsen could confirm that this compound undergoes ligand exchange with dimercaptosuccinic acid resulting in the formation of the well-known Tc(V)-DMSA complex /38/.

Table 7.6. Tc-complexes with aminopolycarbonic acid ligands

chemically characterized

Tc(V)	
TcO $EDTA^-$	
Tc(IV)	
H_6 TcO DTPA	yellow compound
H_6 $Tc_2O(DTPA)_2 \cdot 2\, H_2O$	brown compound
$(NH_4)_4$ H_2 $Tc_2O(DTPA)_2 \cdot 4\, H_2O$	brown compound
H_2 TcO DTPA $\cdot$ 3 H_2O	yellow compound
HEDTA TcO_2Tc HEDTA $\cdot$ 5 H_2O	red compound
K_2 $Tc(OH)_2$ NTA_2	red compound
K_3H_3 TcO(OH) $NTA_3 \cdot 6\, H_2O$	brown compound
Tc(III)	
Tc $(HIDA)_2^-$	

7. Technetium chelates with arsines and phosphines

Shortly only the Tc(III)-chelates with substituted arsines and phosphines should be mentioned. This chemically comprehensive characterized group studied by Fergusson /8, 49/ thirty years ago was introduced into the radiopharmaceutical evaluation by Deutsch /50/ recently.
The results so far obtained imply two *conclusions* which are important for the design of bioinorganic technetium chelates.

1. The in vivo stability of a technetium chelate is determined by its kinetic inertness because the chemical constituents of the organism dispose of the same donors but frequently at higher concentrations.
2. Donor sites and biologically interactive molecular regions of biological substances and drugs are often identical. Therefore, technetium labelling can obstruct the biospecificity of the ligand.

8. Biodistribution and rationales for radiopharmaceutical design

Reliable experimental findings concerning the relationships between the chemical nature of metal chelates and their biodistribution are rather scarce. Therefore, considerations about approaches to a rational designing of technetium radiopharmaceuticals are more or less speculative at present.
Apart from the trivial cases of organ imaging by enhanced vascularization, particle trapping and colloid sequestration, an organ-specific technetium chelate must fulfil two principal requirements:
At first it must reach the target-tissue and, secondly, it must be immobilized or at least retained in the target tissue for a certain period.
With respect to the first question we can refer to the pharmacological rules of appropriate physico-chemical properties for drug transport. A suitable size and charge of the pharmaceutical determines the distribution space, e.g. plasma space and extracellular fluid. The biokinetics is strongly influenced by the protein binding of the substance and can be varied by altering the lipophilicity. The excretion pattern depends mainly on the hydrophilic or lipophilic character. Hydrophilic drugs prefer the renal pathway, and follow, if they are anions, the tubular sezernation in certain cases. Lipophilic substances prefer the hepatobiliar way and will be excreted actively, if they can imitate the endogeneous bilirubin.
Technetium chelates which are provided to be localized within the cell must be lipophilic and neutral in order to be able to penetrate the cell membrane by passive diffusion.
If cationic technetium chelates can be actively transported by amine or potassium pumps, as discussed occasionally /50/ remains speculative until experimental evidence is provided.
Many papers in the current literature (e.g. /1, 51–53/) evidence the efficient application of these rules to optimizing osteotropic, hepatobiliary and other agents.

The requirements which have to be met under the second condition are much more obscure. Classic pharmacology has shown that the interactions with biological structures are extremely specific with regard to certain stereo-chemical molecular arrangements. Commonly labelling by technetium alters the molecular structure and obstructs the functional groups by complex formation. From this point of view, the systematical design of bioinorganic analogues of biointeractive substances endogeneous or drugs, has only a minor chance. But nevertheless it is not hopeless as the casually found hepatobiliary agents, which are similar to bilirubin, and more recently the Tc-dads /53/ compound, which behaves similar to hippurane, demonstrate.

If we refer to the most popular interpretation of the bone uptake of osteotropic agents, the interaction between the external functional groups of a technetium chelate with the biostructure, in this case the Ca-apatite of the growing bones, can be considered to be another approach.

In this respect it is interesting to recognize that also technetium chelates with nonphosphorous ligands like meso Tc(V)-DMSA, Tc(IV)DMSA, Tc(IV)-tartrate and several polymeric Tc(IV)-chelates with hydroxycarbonic acids exhibit pronounced osteotropic properties (Table 7.7.). Probably an important condition for bone uptake is the presence of at least three calcium affine oxygen ligator atoms situated in the same plane at distances of about 0.3 nm.

Therefore, the design of in vivo stable technetium chelates, which have free external functional groups in an appropriate stereo-chemical arrangement meeting the biointeractive regions, could be a further useful rationale.

Reactivity with regard to ligand exchange seems to be an additional principle of localizing technetium chelates in certain organs. We have found that the kidney uptake of the technetium hydroxycarbonic acid chelates is confined to the Tc(V)-oxidation state regardless of the chemical structure of these compounds (Table 7.8.). The only common chemical property in this group is the pronounced tendency to exchange hydroxycarbonic acid ligands for thiol compounds. The half-life of the exchange occurs within minutes.

Table 7.7. Organ distribution, wistar-rats; 60 min p. i.

Preparation	% injected dose skeleton [1]	urine	blood [2]
^{99m}Tc-HEDP	48	43	0.11
^{99m}Tc-MDP	52	40	0.14
^{99}Tc(V)-DMSA (meso)	38	48	0.37
^{99}Tc-citrate (brown)	26	46	0.82
^{99}Tc(IV)-tartrate	24	57	0.41
^{99m}Tc-tartrate	16	43	0.38
^{99}Tc-tartrate (brown)	15	26	—
^{99}Tc-tartrate (violet)	18	50	—

1 % injected dose (femur) × 20

2 % injected dose/gram

Table 7.8.

Complex compound	Tc	Kidney uptake % D/g	Femur uptake % D	Tc	Kidney uptake % D/g	Femur uptake % D
Citrate	V	8.6	0.3	IV	2.9	0.4
Malate	V	13.3	0.2	IV	2.7	0.4
Tartrate	V	13.2	0.2	IV	1.7	1.2
α-hydroxyisobutyrate	V	8.3	0.3	IV	1.5	0.3
Gluconate	V	9.9	0.1	IV	1.1	0.4

We suppose that such a process with SH-proteins can be regarded as the localizing mechanism in organs or tissues which are rich in such proteins, e.g. the renal cortes but also infarctions and necrotic tumours. Taking into consideration the very interesting results obtained by Deutsch et al. /50/ with many chemically very similar Tc(III)-arsine complexes having different auxiliar ligands, it might be possible that the pronounced heart uptake of only some of these can be caused by an exceptional reacivity of these particular compounds.
The design of reactive chelates possessing appropriate physico-chemical properties could be a further successful rationale. In this respect the preservation of the reactive oxidation state in Tc-99m preparations would be challenge to further investigations.

References

1. *Loberg, M. D., M. Cooper, E. Harvey, P. Callary, W. Faith:* J. Nucl. Med. **17** (1976), 633.
2. *Eckelman, W. C.:* J. Label. Comp. **XIII** (1977), 2.
3. *Deutsch, E., W. R. Heineman, R. Hurst, J. C. Sullivan, W. A. Mulac, S. Gordon:* J. Chem. Soc. Chem. Comm. (1978), 1038.
4. *Johannsen, B., H. Spies:* Chemie und Radiopharmakologie von Technetiumverbindungen. Akademie der Wissenschaften der DDR, 1981, p. 169.
5. *DePamphilis, B. V., A. G. Jones, M. A. Davies, A. Davison:* J. Amer. Chem. Soc. **100** (1978), 5570.
6. *Deutsch, E., R. C. Elder, B. A. Lange, M. J. Vaal, D. G. Lay:* Proc. Natl. acad. Sci. (USA) **73** (1975), 4287.
7. *Deutsch, E.:* Proc. 2nd Intern. Symp. Radiopharm. Seattle 1979, p. 129.
8. *Fergusson, J. E., F. S. Nyholm:* Nature **183** (1959), 10—39.
9. *Hambrught, P., J. McRae, P. E. Valk, A. J. Bearden, B. A. Shipley:* J. Nucl. Med. **16** (1975), 478.
10. *Johannsen, B., R. Syhre, H. Spies, R. Muenze:* J. Nucl. Med. **19** (1978), 816.
11. *Johannsen, B., H. Spies, R. Syhre:* Radiochem. Radioanal. Lett **30** (1978), 111.
12. *Kusina, A. F., A. A. Oblova, W. I. Spitzin:* Z. anorg. Chem. (russ.) **XVII** (1972), 2620.
13. *Loberg, M. D., A. T. Fields:* Intern. J. appl. Rad. Isotopes **29** (1978), 167.
14. *Mazzi, U., G. DePaoli, O. Traverso:* J. inorg. nucl. Chem. **39** (1977), 1090.
15. *Mazzi, U., G. DePaoli, D. DeBernardo, L. Magon:* J. inorg. nucl. Chem. **38** (1976), 721.

16. *Muenze, R.:* Radiochem. Radioanal. Lett. **20** (1977), 61.
17. *Muenze, R., B. Grossmann:* Radiochem. Radioanal. Lett. **43** (1980), 215.
18. *Noll, B., S. Seifert, R. Muenze:* Radiochem. Radioanal. Lett. **43** (1980), 215.
19. *Schwochau, K., L. Astheimer, J. Hauk, H. J. Schenk:* Angewandte Chemie **86** (1974), 350.
20. *Smith, J. E., E. F. Byrne, F. A. Cotton, J. C. Sekutowski:* J. Amer. Chem. Soc. **100** (1978), 5571.
21. *Spies, H., B. Johannsen:* Inorg. Chim. Acta **33** (1979), L 133.
22. *Spies, H., B. Johannsen, R. Muenze:* Z. Chem. **20** (1980), 222.
23. *Steigman, J., L. Hwang, S. Srivastava:* J. Label. Comp. Radiopharm. **13** (1977), 160; 162.
24. *Kountz, S. N., I. Yeh, R. Wood, I. Cohen, P. Kiss:* Nature **218** (1967), 1397.
25. *Harper, P. V., K. A. Lathrop, G. M. Hinn* et al.: Radioactive Pharmaceuticals. AEC Symp. Ser. 6 (1966), 347.
26. *Steigman, J., Meinken, P. Richards:* Int. J. appl. Radiat. Isotopes **26** (1975), 601.
27. *Muenze, R., I. Hoffmann:* Radiochem. Radioanal. Lett. **48** (1981), 289.
28. *Muenze, R., I. Hoffmann, R. Dreyer, I. Dreyer:* J. Radioanal. Chem. 1982 accepted.
29. *Kievits, W. D.:* J. Label. Comp. Radiopharm. **10** (1981), 136.
30. *Muenze, R.:* Radiochem. Radioanal. Lett. **31** (1977), 51.
31. *Muenze, R.:* Radiochem. Radioanal. Lett. **48** (1981), 281.
32. *Russell, C. D., A. C. Cash:* Intern. J. appl. Radiat. Isotopes **30** (1979), 483.
33. *Muenze, R.:* Radiochem. Radioanal. Lett. **30** (1979), 117.
34. *Byrne, E. F., Smith, J. E.:* Inorg. Chem. **19** (1979), 1832.
35. *Hartmann, H., D. Mocker, R. Muenze:* unpublished.
36. *Troutner, D. E., J. Simon, A. R. Ketring, W. A. Volkert, R. A. Holmes:* J. Nucl. Med. **21** (1980), 443.
37. *Thornback, J. R., A. E. Theobald:* Intern. J. appl. Radiat. Isotopes **32** (1981), 838.
38. *Johannsen, B.:* Intern. J. appl. Radiat. Isotopes **33** (1982), 429.
39. *Seifert, S., R. Muenze, B. Johannsen:* submitted to Radiochem. Radioanal. Lett. 1982.
40. *Russell, C. D., A. G. Cash:* J. Nucl. Med. **20** (1979), 532.
41. *Deutsch, E.:* cit. in: *G. Bandoli, U. Mazzi, E. Roncari, E. Deutsch:* Crystal Structure of Technetium Compounds (review).
42. *Bürgi, H. B., G. Anderegg, P. Blaeuenstein:* Inorg. Chem. **20** (1981), 3864.
43. *Seifert, S., B. Noll, R. Muenze:* Intern. J. appl. Radiat. Isotopes 33 (1982) 1393.
44. *Noll, B., S. Seifert, R. Muenze:* Intern. J. appl. Radiat. Isotopes 34 (1983) 581.
45. *Terry, A. A., H. E. Zittel:* Anal. Chem. **35** (1963), 614.
46. *Russell, C. D., A. G. Cash:* Intern. J. appl. Radiat. Isotopes **30** (1979), 485.
47. *Libson, K., E. A. Deutsch, B. L. Barnett:* J. Amer. Chem. Soc. **102** (1980), 2476.
48. *Pinkerton, T. C., W. R. Heineman, E. Deutsch:* Anal. Chem. **52** (1980), 1106.
49. *Fergusson, J. E., R. S. Nyholm:* Chem. and Ind. (1960), 347.
50. *Deutsch, E., K. A. Glavan, I. J. Sodd, H. Nishiyama, P. L. Ferguson:* J. Nucl. Med. **22** (1981), 897.
51. *Mahlstedt, J., K. Joseph:* Nuc compact **10** (1979), 147.
52. *Wistow, B., G. Subramanian, J. G. McAfee, R. Hall, G. Gaque, R. Henderson:* J. Nucl. Med. **17** (1976), 545.
53. *Burns, D., L. Marzilli, D. Sowa, D. Baum, H. N. Wagner:* J. Nucl. Med. **18** (1977), 624.

Addendum

Very recent developments in technetium chemistry /1/ are summarized as follows:

Tc^{V} compounds with Tc-N core

Based on the synthesis of the first Tc-N compound by *Wahren* et al *Baldas* and *Bonnyman* elaborated syntheses for a manifold of different Tc-N compounds at the carrier (^{99}Tc) as well as at the non carrier added (^{99m}Tc) level. Ligands include many of these which proved to be radiopharmaceutically relevant with Tc-O compounds. The chemistry of the Tc-N compounds differs from that of the Tc-O compounds by a considerably greater stability of the Tc-N bond which yields on the other hand lower stability for the attached ligands. These peculiarities result in a different biodistribution and biokinetics than for the well known Tc^{V}-O compounds.
The Tc-N complexes can be categorized according to their retention in the blood pool into three groups:

1. Tc-N compounds with ligands like tropolone, acetylacetone and mercaptopyrimidin are retained by cellular constituents of the blood and show a very slow clearance.
2. Tc-N compounds with EHDP, MDP, HIDA, DTPA, citrate etc. exhibit moderate blood pool activity which results from ligand exchange with the blood proteins.
3. Finally, many Tc-N compounds with thiol ligands such as thio-oxime, penicillamine, thioguanine, DMSA, diethyldithiocarbamate thiomalate, mercaptopurine, gluthathion and cysteine are sufficiently stable in vivo for clearance via the hepatobiliary or renal pathway according the lipophilicity of the compound.

Tc^{V} compounds suitable for brain imaging

By appropriate substitution bis-amino-dithiol as well as amine-oxime complexes were obtained which are sufficiently long retained in the brain for single photon studies (SPECT). On approach resembles the first blood/brain barrier penetrating compound found by Kung which is rapidly extracted from the blood stream, however immediately after first passage washed out. By shifting the basic piperydyl substituent from the carbon atom to the neighboured nitrogen (Fig. 7.7.) a neutral lipophilic technetium complex was obtained which showed a half life in the brain of monkeys and baboons around 60 minutes which permits SPECT studies.
Even superior to that are the hexamethylated propene diamine oxime complex which remains practically constant in the human brain and yields perfusion tomograms of excellent technical quality.

Novel myocard affine technetium complexes

Novel cationic technetium compounds have been synthesized and evaluated as potential myocardial agents in different animals and in some cases also in men. *Deutsch* found a striking difference in the biodistribution of $Tc^{III}Cl_2$ $DMPE_2^+$ and the chemically very similar rhenium compound $Re^{III}Cl_2DMPE_2^+$ which was interpreted as an in vivo reduction of the technetium compound.

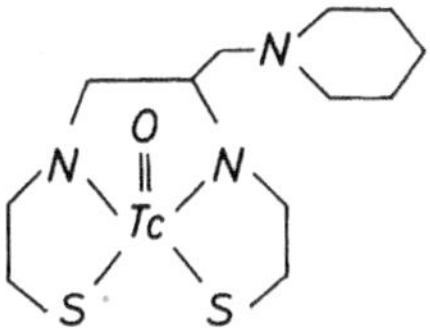

Kung's compound
not persisting
brain uptake

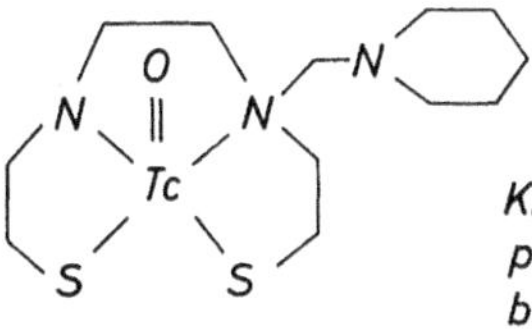

Kramer's compound
persisting
brain uptake

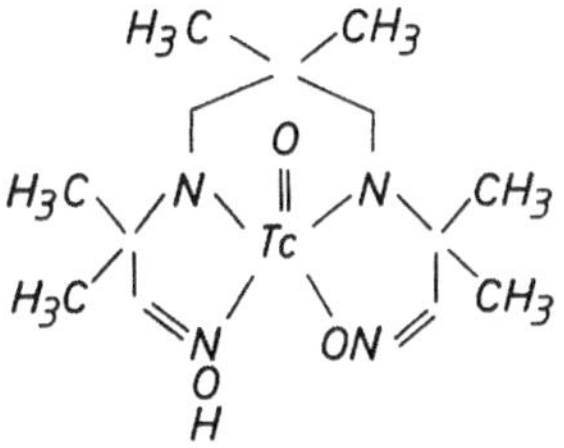

Tc-HM Pn A O

Fig. 7.7.

The rhenium complex cannot be reduced in the biological available potential range. In conclusion more stable Tc^{III} cationic complexes were sought which should not be washed out as $Tc^{III}Cl_2DMPE_2^+$ does. By reduction of the Tc^V-oxo complex with the dinegative acetylacetonethylenediamine *Schiff*-base as the ligand with tertiary alkylphosphines unipositive mixed ligand complexes were obtained. One of them, the $Tc^{III}(acac_2en)(P(CH_3)_3)_2^+$ (Fig. 7.8) yielded excellent images of the myocardium of different animals which justifies an evaluation in man.

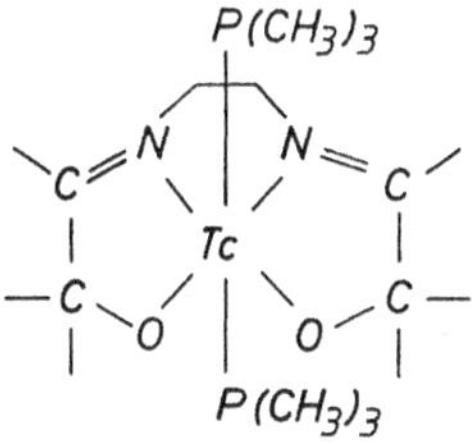

$[Tc^{III}(acac_2en)(P(CH_3)_3)_2]^+$

Fig. 7.8.

In the Tc^{I} field the most interesting result was the evaluation of the cationic Tc^{I}-hexakis(t-butylisonitrile) complex (Tc-TBIN) in man which revealed that this compound is the only one of the hitherto known Tc^{I} complexes which shows similar biodistribution and kinetics in animals and man. Further novel Tc^{I} complexes with phosphites such as Tc^{I} hexakistrimethylphosphite (Tc-TMP) and the Tc^{I} tris(bismethoxy(phosphino)ethane) (Tc POM-POM) were synthesized and evaluated in animals and man. Despite promising biodistribution the clinical studies were disappointing. The general species dependence of these compounds resulted in a rather slow blood clearance.

An interesting novel approach was presented by *Dean* et al. who succeded in the non carrier added synthesis of the unipositive dibenzene-technetium-(I) cation by heating dry pertechnetate with benzene in the presence of aluminium and aluminiumchloride at 135° centigrade for 1 hour. The compound is sufficiently stable in physiological medium and distinguished by an extremely low protein binding (11 %). Heart uptake and heart/liver as well as heart/blood ratios are much to low. However, in the series of methylated benzene ligand members with very promising parameters were found in the range of tetra-hexamethylenebenzene. These compounds yielded excellent myocard images in animals, however conclusions with regard to eventual clinical utility have to expect evaluation in man.

Reference

1. Presentations at the II. International Symposium on Technetium in Chemistry and Medicine held at Padua 9–11. Sept. 1985 Proceedings in press.

Chapter 8

Technetium-99m Radiopharmaceuticals: Their Chemical Potential and Limitations

Bernd Johannsen

1. Introduction

As pointed out in Chapter 13, the principle of preparing Tc-99m radiopharmaceuticals is that of complex formation. Virtually Tc-complexes are stable compounds; it is thus the overall structure of the complex which determines the biodistribution of the tracer.

In biological systems, technetium complexes are by nature foreign substances made from an artificial element. Because of this striking feature, predominance of merely unspecific distribution and excretion rather than "specific" accumulation or even metabolism is the unpleasant consequence, unless there is a profitable reactivity of the coordinate ligands or the complex could be able, by chance, to mimic metabolic substrates, etc. (such as TcO_4^- with respect to the iodide pump) due to similar molecular parameters (see analogue concept, p. 169). Depending on the process to be traced, the requirements concerning the chemical structure or properties may be quite different, ranging from simple physico-chemical properties (e.g. size, lipophylicity) to extremely subtle molecular parameters.

Anyway, a proper understanding of the composition, structure and properties of Tc-compounds will have to be acquired.

2. Need for model studies with long-lived Tc-99

Thanks to the Mo-99/Tc-99m generator, Tc-99m is readily available in neutral aqueous solution. The total "no-carrier-added" concentration of both Tc-99m and Tc-99 in generator eluates is only 10^{-7} mol · l^{-1} or less. Therefore, we are faced with the problem that the very low concentration of technetium prevents the use of conventional analytical methods such as elemental analysis, mass spectroscopy, I.R., VIS, NMR and ESR spectroscopy, X-ray crystallography, etc. Analytical procedures hitherto used for the characterization of labelling reactions and products are confined to methods only offering unspecific information on radioactivity distribution patterns. Though very useful for most routine quality controls of Tc-99m radiopharmaceuticals (determination of radiochemical purity, labelling yield), they do not provide the badly needed insight into the exact composition and structure of Tc-species. Apart from the need for new analytical techniques adequate to the "no-carrier-added" level, the general approach to the synthesis and characterization of defined Tc-complexes is the study of macros-

copic quantity of long-lived "carrier" Tc-99 (for lack of stable isotopes). Tc-99, which emits only weak beta particles with a half-life of $2.1 \cdot 10^5$ years, is available in gram quantities as a major fission product. However, the use of carrier-Tc necessarily raises questions, concerning concentration effects, in the interpretation of the results.
Several research groups are devoting considerable effort to achieving a better understanding of basic technetium chemistry, thus closing the gap between clinical application and chemical research.

3. Reduction of TcO_4^-

Tc-99m radiopharmaceuticals are prepared for reasons of simplicity from TcO_4^- in aqueous saline (generator eluate). With the exception of pertechnetate itself (and Tc-sulphur colloid), all other radiopharmaceuticals require the reduction of Tc(VII) to some lower, more reactive oxidation state followed by stabilizing this form of technetium by the ligand involved.
The redox chemistry of Tc has been extensively studied in almost every oxidation state ranging between + 7 and − 1. From the radiopharmaceutical point of view, only Tc(V), Tc(IV) and Tc(III), which occur under the conditions currently used, are of particular interest. Recently also Tc(I) has been considered to be useful in radiotracer design.
Apart from electrochemical reduction, a large number of reducing agents have been used. But stannous chloride and stannous complexes, respectively, are compounds which have been most widely used in radiopharmaceutical preparation. Currently used instant kits are based on stannous/ligand systems.
The oxidation state of technetium in radiopharmaceuticals is not definitely known. Our present understanding of the oxidation state and the core in the complexes has been improved by model studies with Tc-99 (Table 8.1.).
Model Tc(V)-complexes have either one or two oxo groups that remain bonded to Tc. Thus, the presence of a multiply-bonded oxo group (or nitrido group in special cases) is an outstanding feature of technetium(V). Tc(IV) prepared in aqueous NaCl solution is likely to involve oxygen- and/or OH-containing cores. Structurally, Tc(III) resembles Fe(III).

Table 8.1. Oxidation states of Tc and cores of Tc-complexes

Oxidation state	Core
+ VII	TcO_4^- (starting material)
+ V	TcO^{3+}; trans-TcO_2^+; TcN^{2+}
+ IV	$TcO(OH)^+$; $Tc(OH)_2^{2+}$
+ III	Tc^{3+}
+ I	Tc^+

It is expected that the oxidation state of technetium will influence the biokinetics of the corresponding complexes due to different cores, geometries and reactivities.
When using the stannous chloride method in the usual manner (high excess of Sn(II) over Tc), the reaction proceeds very quickly in a two-electrons reduction of Tc(VII) d^0 to Tc(V) d^2 and then to lower oxidation states, if there is no kinetic inhibition. Depending on the reaction conditions, various complexes may occur. The kinetics and chemistry of reduction and complex formation influence radiochemical purity. In order to achieve an optimal performance, each system with or without stannous precursor (kit) would have to be examined separately. Optimum preparation consists in reducing pertechnetate to a given oxidation state for the desired complex formation without side-reactions of either tin or technetium.
As Sn(II) is a very important ingredient, attention has to focus on its role in preparing Tc-99m radiopharmaceuticals including interfering oxidation and hydrolysis. In order to minimize the loss of Sn(II) due to oxidation, inert atmosphere and a stabilizer such as ascorbate are used.

4. Ligands

In general, complex formation occurs for compounds possessing functional groups such as hydroxyl, carboxyl, amino, mercapto, phosphonate, etc. in an appropriate molecular arrangement. Bidentate and multidentate ligands are most commonly used.
The higher oxidation states of technetium prefer ligands which are able to compensate the very high positive charge of the central atom. Therefore, at the oxidation state + 5 technetium is stabilized as oxotechnetium (or nitridotechnetium) with compounds possessing mercapto, amino, hydroxyl, and carboxyl groups. At the lower oxidation states ligands containing soft donor atoms such as sulphur, phosphorus and arsenic ones or ligands, with π-donor properties are preferred. Due to the border-line behaviour of Tc at the oxidation states + 5, + 4 and + 3, there is much room for variance in Tc-binding and thus radiotracer design.
The failure, to "label" certain interesting substances directly, has led to derivatives which are formed by covalent binding of a chelating moiety (DTPA, IDA, etc.) to the molecule of interest. The term "bifunctional" used in this context is, however, misleading.
For reasons of completeness, also Tc-labelled particles, colloids and cells should be mentioned here. The compounds listed in Table 8.2. are meant to exemplify the wide range of Tc-99m radiopharmaceuticals.

5. Complex formation

As complex formation is the underlying principle of preparing Tc-99m radiopharmaceuticals, the preparation as well as the properties of Tc-complexes have to be considered from the viewpoint of coordination chemistry. Tc-complexes relevant to radiopharmaceuticals are in general kinetically inert. Neither dissociation or ligand exchange nor hydrolysis is expected to occur in vivo. The important key to interpreting the behaviour of the complexes is the structure and properties of the entire complexes rather than of the free ligands. As a result of complex formation, the reactivity of the original ligand is altered insofar as functional groups are completely or

Table 8.2. Some typical ligands used for preparing Tc-99m complexes and other substances which can be labelled with Tc-99m

DTPA DMSA glucoheptonate citrate DADS	for kidney agents
HIDA and various analogues (EHIDA, PIPIDA, BIDA, etc.) pyridoxylideneaminates (Schiff bases) pyridoxylaminates	for hepatobiliary agents
diphosphonates (EHDP, MDP, HMDP, DPD) pyrophosphate	for bone agents
DIARS DMPE isonitriles	for heart-imaging agents
albumin urokinase fibrinogen streptokinase plasmin polypeptides antibodies	labelled proteins
erythrocytes leucocytes platelets lymphocytes	labelled cells
macroaggregated albumin microspheres millimicrospheres sulphur colloid antimony sulfid colloid phytate	labelled particles and colloids

partly involved in binding Tc. There is an unavoidable increase in size and more or less alteration of charge. The term "labelling" has veiled this state of affairs.

One of the main questions arising is the exact structure of Tc-99m radiotracers including oxidation state, type of ligands, coordination number, geometry as well as resulting size, charge, complex stability, reactivity, etc. both in vitro and in vivo. Much of our knowledge in this context has been aquired only during the past decade.

The usual radiopharmaceutical synthesis is confined to conversion of TcO_4^- at 10^{-7} mol · l^{-1} or less concentration in aqueous saline without any purification step yielding the radiochemically pure product within 15 minutes ($t_{1/2}$ 1–2 min!). This requirement can be met by stannous ions and ligands which are in high excess over Tc. This implies pseudo first-order kinetics with respect to Tc-concentration. The pertechnetate disappearance rate becomes independent of the TcO_4^--concentration.

Thanks to the very low Tc-concentration, such phenomena as disproportionation, autocatalytic reactions or formation of dimers or oligomers (apart from heteronuclear products in the case of tin or other metals involved) will not be observed during the time concerned.

The above-mentioned restrictions concerning reaction route and rate are not valid for "carrier" Tc-99 experiments at macroscopic concentrations. The use of appropriate synthesis routes makes it possible to selectively prepare different products which proved to be very helpful in understanding technetium chemistry and to provide the basis for urgently needed structure-biodistribution relationships, though we must realize that the compounds may differ from those of "no-carrier-added" Tc-99m. Extrapolation from model Tc-99 compounds to Tc-99m radiopharmaceuticals may be justified, if identical chromatography (preferably HPLC), electrophoretic and biodistribution patterns are proved.

As far as they are relevant to Tc-99m radiopharmaceuticals, aspects of coordination chemistry of technetium at the oxidation states will be briefly discussed.

6. Chemistry of technetium (V)

Formal two-electron reduction of pertechnetate leads to Tc(V) which can be stabilized in the presence of appropriate ligands, if further reduction to lower oxidation states is avoided. As indicated above, the complexes are characterized by either one or two oxo groups that remain bonded to Tc: a) TcO^{3+} core, b) trans-TcO_2^+ core and—provided there is sufficient Tc-concentration— c) $Tc_2O_3^{4+}$ core. The existence of stable Tc(V)-complexes has not been appreciated for a long time. Actually, there is a large group of ligands, mostly symmetrical and unsymmetrical bidentate and tetradentate ligands, capable of stabilizing oxotechnetium(V). O, N, S, As, and P function as donor atoms.

The structure of resulting monomeric complexes is schematically shown in

Figure 8.1. A number of complexes have been fully characterized structurally, among them the five-coordinate complex with four sulphur donor atoms. The oxo group forms the apex of an approximate square pyramid. The four sulphur atoms form the basal plane. The well-investigated class of bis(dithiolato)oxotechnetium(V)-complexes involves the complex with

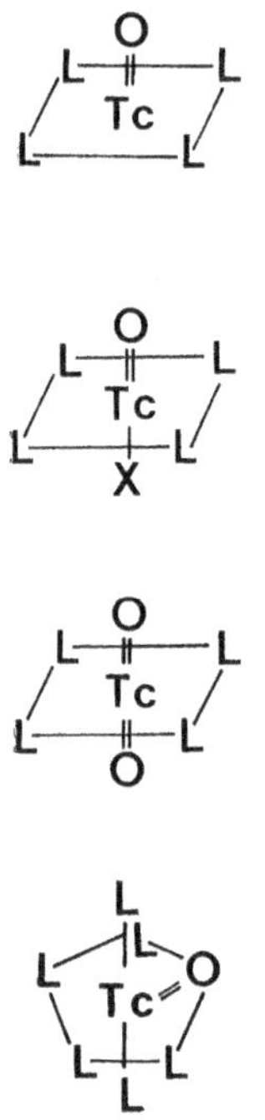

Fig. 8.1. Some types of monomeric Tc(V)-complexes
L = donor atoms of the ligand,
X = Cl, Br, OR

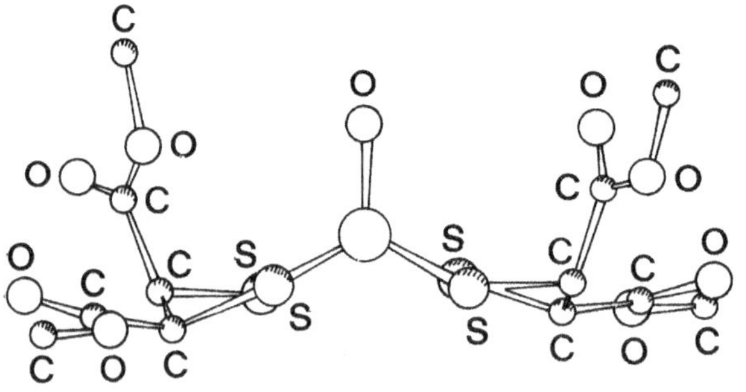

Fig. 8.2. Structure of Tc(V)-complex of meso-dimercaptosuccinic acid dimethyl ester

meso-dimercaptosuccinic acid ester. Its structure is shown in Figure 8.2. It is of great interest that isomers are expected to occur (Figure 8.3.).
O-coordinate complexes can be prepared with hydroxycarboxylates, e.g. gluconate, citrate, α-hydroxyisobutyrate. Proceeding from Tc-99(V)-gluconate which has proved to be a convenient starting material for the synthesis of Tc(V)-thiolato complexes, complexes of N-donor ligands (pyridine, ethylenediamine, etc.) could also be prepared.
The one or two oxo groups in the core appear to be important for neutralizing the high formal charge of Tc(V). If the ancillary ligands donate significantly negative charge to the Tc-centre, only one oxo group (TcO^{3+}-core) is required as in the case of dithiolate ligands. However, if neutral ligands such as pyridine, cyclam or diars are used, two oxo groups are required.

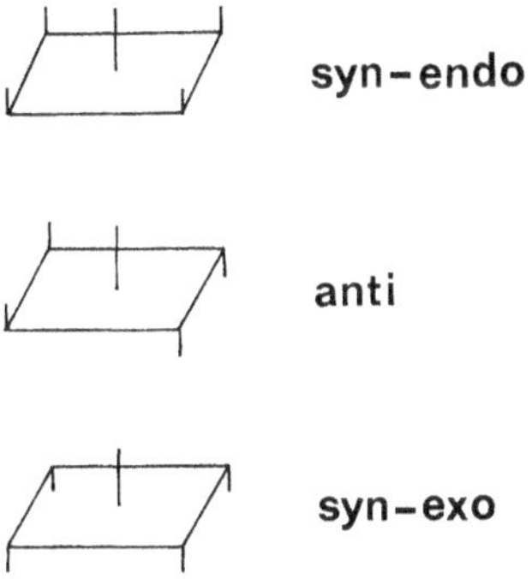

Fig. 8.3. Possible isomers of the Tc(V)-DMSA ester complex shown in Figure 8.2.

As shown in Figure 8.4., the starting O-coordinate complex can be converted into cationic N-coordinate complexes. The cationic N-coordinate Tc(V)-pyridine complex readily undergoes further reaction with dithiols to yield the corresponding Tc-dithiolato complexes. In conclusion, coordination of sulphur donors is favoured in comparison to O and N.
Though such *ligand exchange* reactions of Tc-complexes has not been found to occur in vivo, substitution reactions would be an important point of radiotracer design with respect to specific agents, i.e. agents with a specific behaviour in vivo resulting from an interaction of molecules of the Tc-99m complex and certain molecular counterparts in the biological system. It would be favourable to overcome the inherent substitution inertness of Tc.
For radiopharmaceutical and radiopharmacological purposes Tc(V)-compounds have been used in various respects:

- fundamental radiopharmaceutical studies, e.g. of thiolate complexes,
- search for new renal radiopharmaceuticals (Tc-DADS and derivatives),
- search for cationic and neutral compounds, e.g. complexes of macrocyclic amines (cyclam and derivatives, various N-coordinated complexes, complexes of aminoethanol, etc., nitridotechnetium(V)-complexes, complexes of 8-quinoline and its derivatives, complexes of O, N, S Schiff bases.

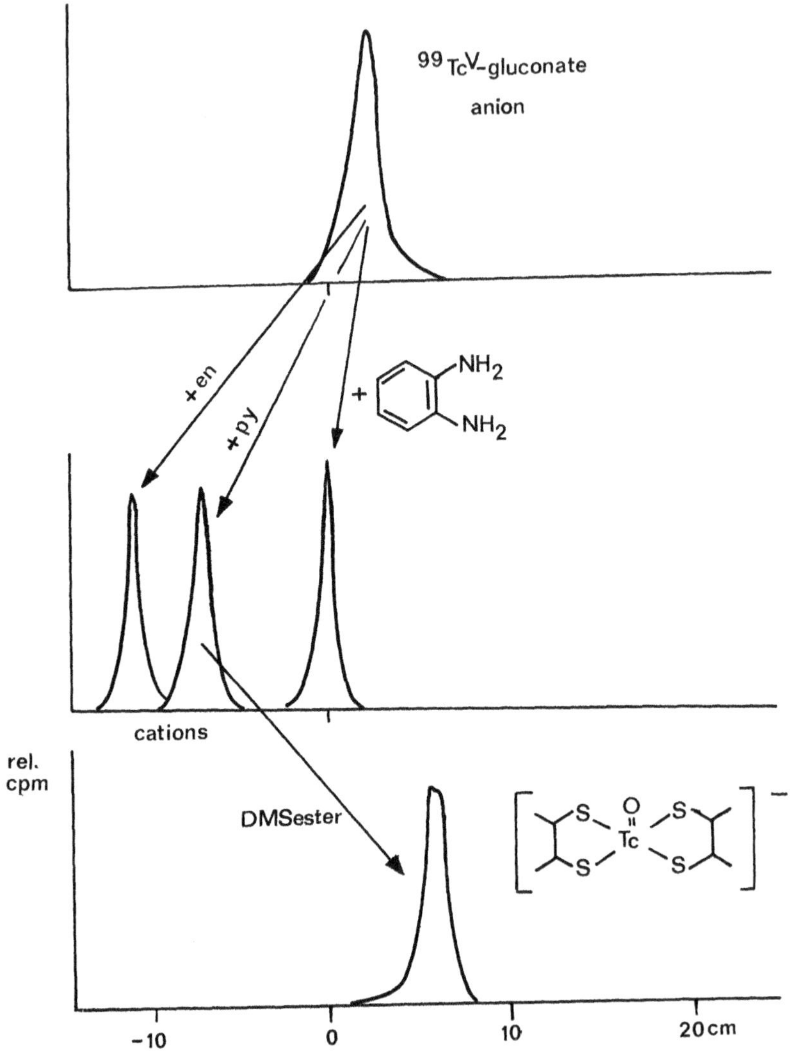

Fig. 8.4. Electrophoretic patterns of the conversion of Tc(V)-gluconate into N-coordinate complexes and subsequent ligand exchange between Tc-pyridine complex and dimercaptosuccinic acid (DMSA) dimethylester

7. Chemistry of technetium (IV)

In strong acid solution, hexahalotechnetates(IV) can be easily obtained by heating pertechnetate in the corresponding halo acid. Salts of $(TcX_6)^{2-}$ are a commonly used starting material for preparing Tc(IV)-model compounds.

The first X-ray structural determination of a Tc-99 complex relevant to a widely used Tc-99m radiopharmaceutical was reported for a product prepared from $(Tc^{IV}Br_6)^{2-}$ by ligand exchange with MDP. The structure of the product with an assumed Tc(IV)-oxidation state consists of infinite polymeric chains. Each technetium bridges two MDP ligands, and each coordinated MDP bridges two technetium atoms. Such a bridge between two technetium centres as postulated for this Tc-99 model polymer $[Tc(MDP)(OH)]_n^-$ is unlikely for the corresponding osteotropic Tc-99m preparation whose composition and structure are still poorly understood. Because of the existence of multiple peaks in HPLC analysis (more than as can be explained by multiple oxidation states of monomeric complexes) and the influence of Tc-concentration, a polymeric nature caused by incorporation of Tc-complexes into Sn-diphosphonate oligomers is assumed.
Thanks to favourable redox potentials and kinetics, pertechnetate can be easily reduced via Tc(V) to Tc(IV) or lower oxidation states. In aqueous solution without appropriate ligands insoluble $TcO_2 \cdot H_2O$ is formed. In the presence of suitable ligands, e.g. bidentate O-donor chelates Tc(IV) complexes are formed. Charakterization of the products has indicated both monomeric and polymeric species, the latter predominantly at higher Tc-concentration. By reducing TcO_4^- by stannous ions in certain systems such as citrate solution, the Tc(V) species may be kinetically trapped. The rate of the Tc(V)/Tc(IV)-reduction step was shown to be proportional to the square root of Tc(IV)-concentration indicating the autocatalytic nature of the process. Because of the very low Tc-concentration in Tc-99m preparations, Tc(V) is assumed to be the predominant oxidation state in this very system. In conclusion, different oxidation states and different molecular structures (monomeric or polymeric) can occur at different Tc-concentration levels existing between macroscopic Tc-99 and "no-carrier-added" Tc-99m solutions.

8. Chemistry of technetium (III) and (I)

The reduction of pertechnetate-99 by excess stannous chloride in a DTPA buffer quickly leads to Tc(III). However, it is difficult to prepare reproducibly and to isolate purely in crystalline form technetium aminopolycarboxylate (NTA, EDTA, DTPA) complexes. With the exception of Tc(V)-EDTA, structural characterization has been rather strongly focussed on their Tc(IV)-compounds which readily undergo oligomerization.
The occurrence of Tc(III) in radiopharmaceuticals has been postulated for the hepatobiliary Tc-HIDA agent. The formulation as six-coordinate anionic 1 : 2 complex is supported by recent studies. Experiments at the millimolar level of Tc-concentration demonstrated similarity of the Tc/HIDA-system to those of Tc-amino-polycarboxylates concerning heterogeneity. Hence, the O,N-donor ligands aminopolycarboxylates are obviously capable of stabilizing Tc(III).
This behaviour is to be taken into account in the case of Tc-"labelling" of DTPA-protein conjugates.

Technetium of low formal charge (Tc(IV)-Tc(I)) is, however, expected to be more favourable stabilized by π-acceptor ligands. π-acceptor ligands stabilize technetium in low oxidation states due to "back donation" of electrons from occupied d-orbitals of the metal to unoccupied orbitals of the ligands. Extensive π-back bonding from low valent technetium occurs in complexes with Tc-As and Tc-P bonds containing arsine, phosphine or phosphonite ligands.

Fundamental studies have been carried out of various phosphine and phosphonite complexes of Tc in the oxidation states + 1, + 2, + 3 and + 4.

The Tc-complex with the diars ligand (diars = o-phenylenebis(dimethylarsine), already prepared three decades ago, was now structurally characterized as Tc(III), which is six-coordinated in a trans-octahedral coordination geometry. Two auxiliary chlorine ligands are situated trans to one another, and the four As atoms from the two nearly coplanar diars ligands form the basal plane of the octahedron. This cationic Tc(III)-complex represents an extremely useful lead structure in the search for myocardial imaging agents.

As far as potential myocardial imaging agents are concerned, bidentate tertiary arsine and phosphine ligands have been used to prepare cationic Tc-(III)-complexes having the formula trans-$[TcD_2X_2]^+$, where D represents the chelating diarsine or diphosphine ligand and X represents Cl, Br, I or pseudohalides. The complexes can be prepared by different routes and from different starting materials. From the radiopharmaceutical point of view, a procedure starting directly from TcO_4^- is most convenient. The ligands mentioned can function both as the reductant and the agent which coordinates to the Tc-core. At present the most promising ligands seems to be ditertiary diphosphine, in particular bis(1,2-dimethylphosphino)ethane (DMPE). Tc-(VII) is successively reduced in excess D to yield well-characterized complexes of Tc(V), Tc(III), Tc(II) or Tc(I), dependent on the reaction conditions (Figure 8.5.). The preparation of myotropic complexes is complicated by the occurrence of the trans-dioxo Tc(V)-complex $[TcD_2O_2]^+$. Proper control of the reaction conditions yields any complex (also at the "no-carrier-added" Tc-99m level) in radiochemical purity higher than 95 %. There are marked species differences in the biodistribution patterns of the $[Tc(DMPE)_2Cl_2]^+$ cation likely to be caused by some serum factors which differ in the species concerned.

The Tc(I) tris complex $[Tc(DMPE)_3]^+$ is also accumulated in the normal myocardium (in a pharmacokinetically different manner compared to the Tc-(III)-DMPE) and recent studies have dealt with this very agent. Further work appears to be necessary in this promising field.

Fundamental chemical research has been devoted to numerous Tc(III)-complexes containing various ligands such as combinations of phosphine, haline, carbonyl, nitrosyl ligands or thiocyanate. The Tc-oxidation state + 3 is not reserved to group V-donor ligands. Recently compounds with thioureas and monothio-β-dicarbonyl ligands have been described.

With isonitriles, a new class of water-soluble complexes of the general formula $[Tc^I(CNR)_6]^+$ can be prepared:

$$[Tc^{III}(tu)_6]^+ \xrightarrow[\text{isonitriles}]{\text{excess of}} [Tc^I(CNR)_6]^+$$

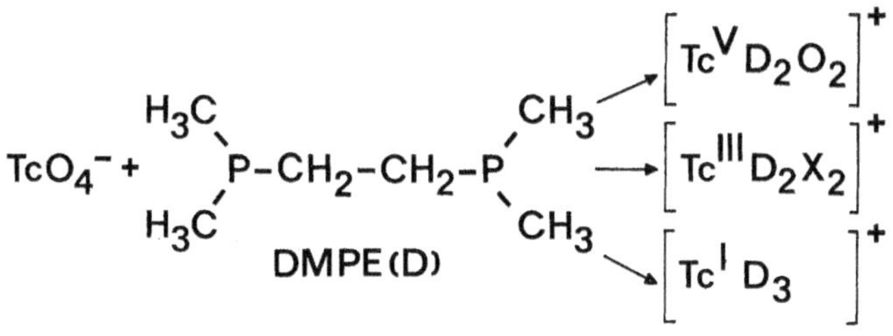

Fig. 8.5. Reaction of pertechnetate with bis(1,2-dimethyl-phosphino)ethane (DMPE) to form myocard affinic Tc(III)- and Tc(I)-complex
X = Cl, Br, I

This class is also obtained by reduction of TcO_4^- in aqueous ethanolic solution containing excess isonitrile ligands.
A number of these cationic complexes (e.g. R = t-butyl) show a promising biological behaviour in animal experiments such as cell labelling and uptake by normal myocardial tissue.
Another radiopharmacologically interesting complex (myocardial accumulation and retention) is the hexakis(trimethylphosphite)-technetium(I)cation.

$$\underset{\text{in methanolic solution}}{NaTcO_4} + P(OCH_3)_3 \xrightarrow[100^\circ C]{30\ min} [Tc(P(OCH_3)_3)_6]^+$$

The monocationic product was characterized as a hexacoordinate octahedral complex. The Tc-99 and Tc-99m preparations were shown to be identical by HPLC.

Selected references (reviews)

1. *Johannsen, B., H. Spies:* Chemie und Radiopharmakologie von Technetiumkomplexen. Akademie der Wissenschaften der DDR, Rossendorf 1981, p. 213.
2. *Srivastava, S. C., P. Richards:* Technetium-labelled compounds. In: Rayudu, G. V. S. (ed.): Radiotracers for medical application. CRC series radiotracers in biology and medicine, CRC Press, Inc. Boca Raton 1983, Vol. I, p. 107.
3. *Deutsch, E., M. Nicolini, H. N. Wagner* jr. (eds.): Technetium in chemistry and nuclear medicine. Cortina International, Verona 1983.
4. *Deutsch, E., K. Libson, S. Jurisson:* Technetium chemistry and technetium radiopharmaceuticals. Progr. Inorg. Chem. **30** (1983), 75.
5. *Schwochau, K.:* The present status of technetium chemistry. Radiochim. Acta **32** (1983), 139.
6. *Münze, R.:* Chemie und Radiopharmakologie organ- bzw. funktionsspezifischer Metallkomplexverbindungen des Technetiums. Isotopenpraxis **19** (1983), 401.
7. *Bandoli, G., U. Mazzi, E. Roncari, E. Deutsch:* Crystal structures of technetium compounds. Coord. Chem. Rev. **44** (1982), 191.

Chapter 9

Chemistry of Indium-Complexes

Ingrid Kaempfer

There are Thakur, Goodwin, Scheffel, McAfee, Welch, Subramanian et al. /1–6/ many reports on the preparation and clinical applications of indium complexes.

In this paper I shall try to give a review of the chemistry of these compounds.

Indium is a metal of the group III A in the periodic classification of elements. In its chemical behaviour indium is similar to its neighbours gallium and thallium, occasionally with cadmium and tin.

Indium has two naturally occurring isotopes—indium-113 (4.33 %) and indium-115 (95.67 %) and over thirty radioactive nuclides. Up to now the isotopes indium-113m and indium-111 have been introduced in nuclear medicine.

In-113m was introduced in 1966 for blood-pool scanning /7/. It has a half-life of 1.7 h and decays by isomeric transition, emitting a 390 keV gamma ray. Indium-113m is commonly obtained from generators. The hydrochloric acid eluate contains In-113m in an almost carrier-free state. Because of its short half-life preparation of indium-113m radiopharmaceuticals requires simple procedures.

For longer investigations indium-111 is the nuclide of choice. Hunter et al. /8/ first reported on the in vivo-application of In-111-chloride as a tumour-localizing agent in 1969.

It is a cyclotron-produced radioisotope, available as the trichloride in 0.05 n HCl. It is produced carrier-free. Indium-111 decays by electron capture with a physical half-life of 67.5 h. It emits two gamma rays at 173 keV (89 %) and 247 keV (94 %). In-111 has ideal physical characteristics for scanning up to one week after administration with a minimum of radiation exposure to the patient.

As a member of group III A in the periodic classification of elements, indium has three possible oxidation states — + 1, + 2 and + 3. The trivalent state dominates in the chemistry of indium. In aqueous solution the trivalent state is also the stablest one. In acidic solution the trivalent indium is the form of hydrated ion.

Above pH 3.4 indium tends to precipitate as hydroxide. The solubility of indiumhydroxide in water is very little. With a large excess of aqueous alkali hydroxide forms a transparent colloidal solution.

The precipitation of $In(OH)_3$ takes place at various pH. It depends on the addition of alkali metal halide, acetate, tartrate and others, to the alkaline solution, which can increase the precipitation pH considerably. This may be due to the formation of indium complexes which will be discussed later.

If we intend to deal with indium complexes in greater detail, we should explain some definitions for better understanding. A complex has been defined as a species formed by the association of a metal ion (acceptor) and a simple inorganic ion or organic molecule (donor atom).
Some ligands are attached to the metal atom by more than one donor atom in such a manner as to form a heterocyclic ring. This type of ring has been given a special name–chelate ring–and the molecule or ion, which it is formed from, is known as a chelating agent. The process of forming a chelate ring is known as chelation. The first chelating molecules discovered were those with two donor atoms, and it was the mode of attachment of the molecules to the metal atom that led Morgan and Drew in 1920 to suggest the name chelate, which is derived from the Greek word η' xηγη' meaning a lobster's claw.
The chelating agents are referred to as a bidentate or polydentate ligand.
Description according to the plan (Figure 9.1.).
A special kind of metal chelates is called "inner complex metallic salts", e.g. compounds formed from metal ion and 8-hydroxyquinoline (oxine) or acetylacetone. They are electrically neutral metal chelates. The inner complexes are notable for their low solubility in water.
After we have dealt with the general terms of the chemistry of complexes, we shall now turn to the interesting indium compounds. Under normal circumstances no oxidation or reduction of the metal ion has been necessary before complex formation.
The metal coordination number in indium compounds can be 4, 5 or 6 but the coordination number six is the most important one.
Indium forms complexes and no chelates with inorganic ions like halides and others.
Some typical ligands commonly used in the formation of organometallics are listed in the Table 9.1.

Table 9.1. Typical ligands of organometallics

Donor Atom	Ligands
N	RNH_2, R_2NH, R_3N, $R_2C{=}NOH$, $R_2C{=}NH$
O	ROH, RCOOH, $R_2C{=}O$
P	RPO_3H
S	RSO_3H, RSH, $R_2C{=}S$

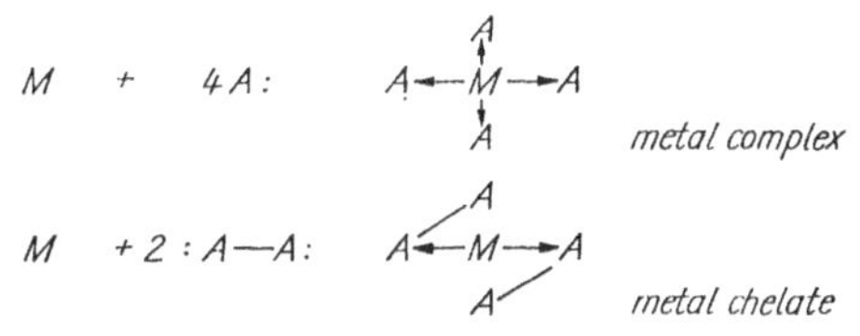

Fig. 9.1. Schema of formation for metal complexes and metal chelates

The chelating agents in the organometallics used for in vivo-investigations should perform several assumptions according to Thakur.
First, they should have a suitable functional group or groups in their molecule, which would form a chelate with indium;
second, they should keep their biological activity unaltered even after chelate formation;
third, they should have specifity to the target tissue; and
fourth, the remainder of the radiopharmaceutical should leave the body through excretion.
A lot of indium radiopharmaceuticals are known.
The stability constants of the complexes are very important parameters. The knowledge of stability constants has contributed much to rationalizing our understanding of the behaviour of metal chelates in solutions, especially chelates in the human body. The high values for stability constants are of importance to the in vivo-stability of compounds.
That means indium chloride can be easy converted into a whole series of indium compounds which have higher values for stability constants. The in vivo-conversion of indium chloride to indium transferrin is established. The known stability constants for organic logards are for the most part much higher than for inorganic ions.
The stability constants of the indium complexes from acetate, oxalate, citrate, tartrate, and formate range also from 10^3 to 10^8. Indium citrate is essentially the same compound as indium chloride from a chemist's point of view.
These complexes also have a weak thermodynamic stability. Therefore the in vivo-stability in these compounds is very little and can be easily converted into several chelates with a greater stability. For that reason they have no specifity.
For instance, when these organometallic compounds are injected, they meet with other chelating agents, e.g. transferrin. As indium transferrin has a greater stability than the other complexes, they can be easily converted into indium transferrin under loss of specifity of the original compounds.

Table 9.2. The stability constants of inorganic indium complexes are generally about 10^3

Compounds	Stability constants ($\log K_1$)
In-111-chloride	2.44
In-111-acetate	3.5
In-111-oxalate	5.3
In-111-citrate	6.18
In-111-acetylacetonate	8
In-111-oxine	10
In-111-nitrilotriacetic acid (NTA)	15.9
In-111-ethylenediaminetriacetic acid (EDTA)	25.3
In-111-diethylenetriaminepentacetic acid (DTPA)	28.4
In-111-transferrin	30

The behaviour of indium bleomycin is similar. Bleomycin is a pharmaceutical with cytostatic properties and therapeutic effects. Several chelating groups are in the bleomycin molecule. In 1973 Merrick showed that trivalent indium chelates with bleomycin. The thermodynamic stability is unknown. But Thakur et al. have reported that 4 h after application of indium bleomycin to a patient, 95 % of the circulation activity is associated with transferrin. However, 5 min after application more than 95 % are still bleomycin-bound. For bleomycin a certain tumour specifity was also found. But until now all the experiments have not shown whether the activity accumulated in the tumour cells is due to labelled transferrin, labelled bleomycin or both.

All results have shown that indium bleomycin has a moderate stability. In the case of chelates which are formed with nitrogen-containing agents, especially EDTA and DTPA, the stability constants are of the order of 10^{20} or higher.

The multidentate ligand diethylenetriaminepentacetic acid forms a strong 1 : 1 chelate with indium. The thermodynamic stability is higher than 10^{20}. This strong chelate has been mainly used for cisternography.

Therefore, none of the chelates with stability constants smaller than 10^{20} has been found to be specific.

Experiments to increase the thermodynamic stability, thereby leading to an increase in specifity, are necessary.

1. Modification of polyaminocarboxylic acids (e.g. EDTA) by reaction with a benzenediazonium salt or by substitution of methylene phosphoric acid instead of acetic acid in the EDTA molecule.
2. Application of lipid-soluble chelates of indium-111 to blood cell labelling (e.g.) erythrocytes, leucocytes and platelets.

The results are promising.

Subramanian /12/ has tried to prepare new chelates on the basis of the polyaminoacetic acids like EDTA, DTPA and nitrilotriacetic acid (NTA). He has substituted methylene phosphoric acid for acetic acid groups in EDTA, DTPA or NTA. These new chelates have retained their strongly chelating properties like EDTA and the others but the new group gains additional bone-seeking properties. The new compounds form chelates with indium. These compounds are strong chelates and, therefore, the in vivo-stability is excellent (Figure 9.2.).

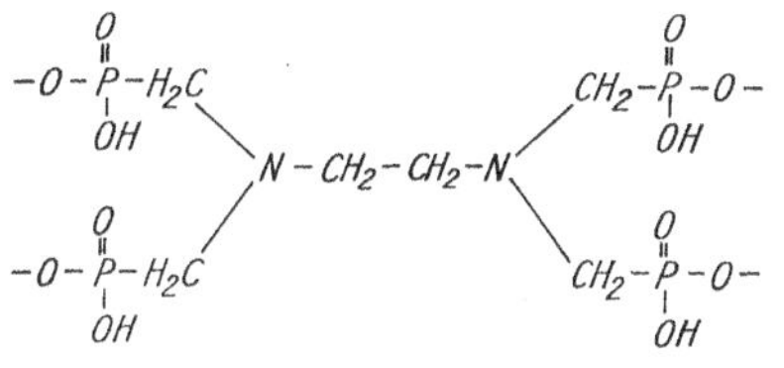

EDTMP

Fig. 9.2. Ethylenediaminetetra (methylene phosphoric) acid (EDTMP)

Another possibility for increasing the thermodynamic stability has been shown by Goodwin /13/. Goodwin et al. have also modified the EDTA molecule by reaction with a benzenediazonium salt. This EDTA-benzenediazonium compound retains its initial metalbinding capability and gains additional ability to link other compounds like proteins and polypeptides. The new-formed indium chelate has an excellent in vitro- and in vivo-stability.
Several compounds such as bovine fibrinogen, human serum albumin and bleomycin were conjugated with indium labelled benzenediazonium-EDTA-chelate, and their distribution was studied. The results are promising.
Another factor, which may influence the specifity of indium chelates, is the lipid solubility of compounds. Lipid solubility is related to the functional groups in a compound. Hydrophobic groups like aryl-, alkyl substituents or hydrocarbon chains enhance the lipid solubility of compounds.
The lipid solubility of a substance is determined by the oil/water partition coefficient.
Also the what are called "inner complex metallic salts" are lipid-soluble. Such chelates are compounds formed from indium ion and acetylacetone, oxine, and tropolone. Lipid-soluble chelates are capable of penetrating lipid cell membranes.
McAfee and Thakur /14/ have surveyed several radioactive agents for labelling phagocytic leucocytes. Water-soluble complexes have no significant cell binding. High labelling efficiency was obtained only with nonpolar complexes which are capable of penetrating lipid cell membranes.
The most promising lipid-soluble agents labelled with indium are oxine, acetylacetone, tropolone and tetraphenylporphyrin. As this field of labelled blood cell components is very important for medicine, I shall describe these compounds:

Indiumoxine

8-hydroxyquinoline is a bidentate ligand. Three molecules of oxine chelate one atom of indium. The thermodynamic stability is moderate; it was estimated at 10^{10}. As we know, compounds with a stability constant of 10^{10} do not possess specifity. Thakur et al. /15/ have shown that indiumoxine added to blood as a whole forms indium transferrin. This compound is also not selective for labelling blood elements. Therefore, it is necessary for labelling isolated specific blood elements. Then it is possible to label red blood cells, platelets, leucocytes as lymphocytes with indiumoxine as well.
The mechanism of labelling has been explored in the case of human neutrophils /16/. It results in an association of the label with the cell elements. Indium can be transferred through the plasma membrane as a lipid-soluble chelate with oxine, and then into the cell. Within the cell indium dissociates from oxine. The low stability may be the cause for the rapid displacement of the indium from the complex. Then the indium links to components in the cell. Subcellular fractionation experiments and gel filtration indicated that the radioactivity was associated with four components of the cytosol fraction. A similar mechanism and transfer of indium oxine to subcellular platelets has been proposed /17/. Other authors /18/ labelled platelets with

H-3-oxine. They said that only traces of 0.01 % of H-3-oxine was connected with the platelets after labelling.
This is another evidence of the theory of labelling mechanism too.
The preparation of indiumoxine depends on several factors. Oxine forms water-insoluble precipitates with more than thirty different metals. It is, therefore, very important to use indium without other tracers like iron, cadmium, zinc, lead, copper, etc. /2/. The purification from heavy metal impurities was performed on an anion-exchange column or by extraction with organic solvents. These procedures remove all the contaminants without loss of indium activity.
The In-111-oxine complex is in vitro-stable, and it is possible to store it for a certain period of time. This is a great advantage for the daily work in a nuclear medical department.
If one modifies the oxine molecule by substituting hydrophilic groups like $-SO_3H$, $-NH_2$ or $-OH$ groups, then they are also capable of chelating with indium. But the labelling of isolated blood cells is not possible because the new compounds are not lipid-soluble and, therefore, not able to penetrate the cell membrane.

Indiumacetylacetone

The method for the labelling of red blood cells was at first reported by Sinn /19/ in 1974.
Three molecules of acetylacetone form a chelate with one indium atom. The stability constant is estimated at 10^8. Similar to indiumoxine, it is useful for labelling isolated cellular blood components such as red blood cells, leucocytes and platelets. The labelling mechanism is similar to the above-described one with oxine. Experiments with C-14-acetylacetone /20/ have indicated that indium dissociates from acetylacetone within the cell and links to cell components. Only 2 % of C-14-acetylacetone were connected with the red blood cells after labelling.

Indiumtropolone

Analogous to indiumoxine and indiumacetylacetone, three tropolone molecules form a chelate with indium /21/. The thermodynamic stability corresponds to indiumoxine.

Indiumtetraphenylporphyrin

This compound is useful for the labelling of blood cell components, but the preparation is very difficult /14/.
We compare the three indium chelates. None of the three indium chelates are specific labels for the various blood cells.
The stability constants of these chelates are moderate from 10^8 to 10^{10}. They are capable of labelling isolated specific blood cells under various conditions.
Selective cell labelling requires a careful cell separation. If the greatest care is exercised, a certain degree of damage of the cells is inevitable. Many reports have stated that the functional viability of labelled blood cells with indiumoxine, indiumtropolone and, with restrictions, indiumacetylacetone are similar to Cr-51-labelled cells.

In our opinion, the indiumoxine and indiumtropolone chelates produce the most promising results with the labelling of blood components.
It is always better to use commercially available compounds such as indiumoxine or indiumoxine sulfate, because it is difficult to remove the heavy metals from indium chloride in the laboratory of a nuclear medical clinic.
A great advantage of tropolone is that it is untoxic for blood elements. Unfortunately, indiumtropolone has not been a fabricated radiopharmaceutical up to now.

Conclusions

There are no indium complexes known to occur naturally in the human body.
Most of the indium radiopharmaceuticals have a weak thermodynamic stability and, therefore, no specifity. It is known that such compounds in the body are converted into indium transferrin.
Experiments to increase the thermodynamic stability, thereby leading to an increase in specifity, have been conducted in two directions:

1. Complexes of benzenediazonium-EDTA, which are labelled with indium and conjugated with proteins, possess the desired stability and thus the organspecifity. Another route was the modification of indium-labelled polyaminocarboxylic acids as bone-seeking agents.
2. Application of lipid-soluble chelates of indiumoxine, indiumtropolone or indiumacetylacetone have shown that it is possible to label isolated cellular blood components, e.g. red blood cells, leucocytes, platelets and others.

The cell viability is unaffected.
The labelling of isolated cellular blood components with indium chelates is a new technique with important applications in clinical medicine.
Cells labelled in this way have shown promising results in localization studies and in investigations of cell kinetics.

References

1. *Thakur, M. L.:* Gallium-67 and indium-111 radiopharmaceuticals. Int. J. Appl. Radiat. Isotopes **28** (1971), 183.
2. *Goodwin, D. A., J. T. Bushberg, P. W. Doherty* et al.: Indium-111 labeled autologous platelets for location of vascular thrombi in humans. J. Nucl. Med. **19** (1978), 626.
3. *Scheffel, U.:* Labeling of human platelets with (In-111)-β-hydroxyquinoline. J. Nucl. Med. **20** (1979), 524.
4. *Grossman, Z. D., B. W. Wisto, J. G. McAfee* et al.: Platelets labeled with oxine complexes of Tc-99m and In-111. Part. 2: Localization of experimental induced vascular lesions. J. Nucl. Med. **19** (1978), 488.
5. *Welch, M. J., T. J. Welch:* Radiopharmaceuticals. Ed. by Subramanian, Rhodes, Cooper, and Sodd. Soc. Nucl. Med., New York 1975.
6. *Segal, A. W., P. Deteix, R. Garcia* et al.: Indium-111-labeling of leucocytes. J. Nucl. Med. **19** (1978), 1238.
7. *Stern, H. S., D. A. Goodwin, U. Scheffel, H. N. Wagner:* In-113m for blood-pool and brain scanning. Nucleonics **24** (1966), 57.

8. *Hunter, W. W., H. W. De Cock:* ^{111}Indium for tumor Localization. J. Nucl. Med. **10** (1969), 342.
9. *Merrick, M. V., J. P. Lavender, C. W. Poole, G. P. Stradlin, M. L. Thakur, L. H. Walter:* In-111-labeled bleomycin. Brit. J. of Radiology **48** (1975), 279.
10. *Thakur, M. L., M. V. Merrick, S. W. Gunaseckera:* Some pharmacological aspects of a new radiopharmaceutical, indium-111-bleomycin. Radiopharmaceuticals and labeled compounds **2** (1974), 183 (IAEA Vienna).
11. *Goodwin, D. A., C. H. Song, R. Fiston, P. Martin:* Radiology **108** (1973), 91.
12. *Subramanian, G., J. G. McAfee, M. Rosenstreich, M. Coco:* Indium-113m labeled polyfunctional phosphonates as bone-imaging agents. J. Nucl. Med. **16** (1975), 1080.
13. *Goodwin, D. A., C. F. Meares, C. I. Diamanti, M. W. Sundberg:* Chelates for radiopharmaceutical labeling. World Cong. Nucl. Med., Tokyo/Japan 1974.
14. *McAfee, J. G., M. L. Thakur:* Survey of radioactive agents for in vitro-labeling of phagocyt leucocytes. J. Nucl. Med. **17** (1976), Part I/II, 480.
15. Analogous /1/.
16. *Thakur, M. L., A. W. Segal, L. Louis, H. J. Welch, J. Hopkins, T. J. Peters:* Indium-111-labeled cellular blood components: Mechanism of labeling and intracellular location in human neutrophils. J. Nucl. Med. **18** (1977), 1020.
17. *Joist, H. J., R. K. Baker, M. L. Thakur* et al.: Indium-111-humanplatelets. J. Lab. Clin. Med. **92** (1978), 829.
18. *Mathias, J.*, et al.: Labeling mechanism and localization of indium-111 in human platelets. J. Nucl. Med. **20** (1979), 659.
19. *Sinn, H., P. Eeorgi, J. Clorius, W. Maier-Borst:* Die Markierung von Erythrozyten mit radioaktiven Indiumisotopen. Nucl. Med. **13** (1978), 180.
20. *Kämpfer, I., K. Nissler:* Unveröffentlichte Ergebnisse.
21. *Burke, J. E. T., S. Roath, D. Ackery, H. P. Wyet:* The comparison of 8-hydroxyquinoline, tropolone, and acetylacetone as mediators in the labeling of polymorphnuclear leucocytes with In-111. A functional study. Eur. J. Nucl. Med. **7** (1982), 73.

Chapter 10

Labelling and Analysis of Protein-Based Radiopharmaceuticals

Győző A. Jánoki

In nuclear medicine biologically significant proteins labelled with various radionuclides have a potential for a variety of diagnostic applications. The method chosen for the labelling of these proteins must fulfil the following conditions:

- the immunological and biological activity of the proteins must be retained after labelling,
- labelling must achieve a high specific activity in some cases,
- labelling and purification of labelled proteins should be carried out under conditions as mild as possible.

Proteins and radionuclides must frequently be used in nuclear medicine as listed in Table 10.1. and 10.2.

1. Methods for labelling proteins

Methods for direct iodination of proteins

Introduction of iodine into the protein chain is performed by electrophylic aromatic substitution. The number of iodine atoms per protein molecule designates the degree of iodination. When this is smaller than two, the labelled proteins have the same biological properties as before labelling. When the degree of iodination is higher than two, the labelled protein disappears from blood circulation and accumulates increasingly in the liver due to phagocyte function.

Table 10.1. Selected protein for use in nuclear medicine

Native form:	
Human Serum Albumin	
Transferrin	dynamic vascular and blood pool studies
Fibrinogen	deep-vein thrombosis detection
Immunoglobulin G	(monoclonal or polyclonal) radio-immuno-detection of tumours
Particulate form:	
Macroaggregated albumin (MAA)	pulmonary perfusion scintigraphy
Human albumin microspheres (HAM)	pulmonary perfusion scintigraphy
Microaggregated albumin (MiAA)	liver and spleen scintigraphy
Albumin nanocolloid (ANC)	bone marrow scintigraphy, lymphoscintigraphy

Table 10.2. Selected radionuclides for protein labelling

Nuclide	$t_{1/2}$	Principal γ-energy (keV)
^{51}Cr	27.7 d	325
^{67}Ga	78 h	90, 184, 296
^{77}Br	57 h	245, 521
^{99m}Tc	6 h	140
^{111}In	2.8 d	173, 247
^{113m}In	104 min	390
^{123}I	13 h	159
^{125}I	60 d	27, 35
^{131}I	8.1 d	364

For direct iodination of proteins the following methods are available:

- iodine monochloride methods /1/,
- chloramine-T methods /2/,
- alternative chemical oxidation methods /3/,
- electrolytic iodination methods /4/,
- enzymatic iodination /5/,
- solid phase system using Iodogen™ or Iodobead™ /6, 7/.

Direct iodination methods often have the associated problems of chemical damage to the protein caused by the reagents, which are used in the reaction.

Conjugation-labelling methods for iodination of protein

Conjugation-labelling methods using N-succinimidyl 3-(4-hydroxy-5-^{125}I-iodophenyl) propionate (Bolton and Hunter reagent) /8/ are more complex. The reagent to be conjugated has first to be radioiodinated, then purified and finally reacted with the protein. These methods can be used to label proteins, peptides which do not contain tyrosine and those in which modification of tyrosine by iodination destroys the required biological activity. The N-succinimidyl group of this reagent condenses with free amino groups of proteins to form conjugate in which the radioiodinated phenyl group is covalently linked to the protein via an amide bond.

^{99m}Tc-labelling of proteins

Most of the existing knowledge about technetium-labelling of proteins pertains to HSA. It may be invalid to extrapolate the experience with HSA to other less familiar proteins and polypeptides. Nevertheless, HSA is readily available and in this context may be used as a convenient model protein for investigation. Several known methods for technetium-labelling HSA are reviewed as follows:

- Ascorbate method /9/,
- Ascorbate-plus-Fe method /10/,
- Fe(II) method /11/,
- Electrolysis method /12/,
- Sn(II) method /13/.

All methods known involve the use of a reducing system or a system that could contain the reducing agent. Technetium used as pertechnetate, does not bind to HSA without such a system. Experiments with technetium-labelling of HSA support the idea that Tc(VII)-reduction to lower valency states is a prerequisite for binding technetium to HSA. Protein complexes of reduced technetium tend to hydrolyze, forming insoluble colloid compounds such as technetium dioxide (TcO_2) or to oxidize or disproportion reforming pertechnetate ions.
These two chemical forms of technetium and the high molecular weight artifacts such as protein dimers, polymers or fragments are the most common radiochemical impurities in ^{99m}Tc-labelled protein radiopharmaceuticals. The insoluble oxide and polymers are concentrated in liver, spleen and the RES bone-marrow, while the highly soluble pertechnetate ions are concentrated in thyroid, stomach, etc.

Protein labelling using bifunctional chelating agents

The *in vivo*-stability of radiometal protein complexes is generally low, which is due to the absence of specific metalbinding receptors which form stable bonds. One can achieve an improvement in the stability of such complexes by involving the content of "bifunctional chelates", first introduced in radiopharmaceutical studies by Sundberg et al. /14/. Various chelons and coupling methods have been tried which lead to compounds having higher labelling yield and a superior *in vivo*-stability compared to directly labelled proteins (Table 10.3.). In choosing the coupling reaction and chelon two requirements have to be considered /15/:

1. The chelon must be attached in such a way that the product retains its original biological and metal-binding properties.
2. The chosen metal ion(s) must form chelates with conjugate, which dissociates very slowly or not at all under *in vitro*- and *in vivo*-conditions.

This method of labelling introduces a rather large charged "foreign" group into the molecule in comparison with e.g. conventional radioiodination. Because of the size and charge of the labelling group, it is important to determine whether the biological properties of the labelled molecule are significantly altered by the "foreign" substituent. This question must be answered for each labelled species of molecule by *in vivo*-observation of its behaviour, rate and route of metabolism, compared to the standard radioiodinated derivative.

Table 10.3. Protein-based radiopharmaceuticals have been developed by introducing strong, metal-chelating groups into biologically active molecules

Biologically active molecules	Chelators	Radiometals
HSA	EDTA	^{111}In
Fibrinogen	DTPA	^{67}Ga
Transferrin	Desferrioxamine (DF)	^{99m}Tc
	iminodiacetic (IDA)	^{67}Cu
Antibodies (or fragments)	di-thiosemicalasone (DTS)	^{97}Ru

2. Analysis of protein compounds labelled with various radioisotopes

The radiopharmaceutical quality of radiolabelled protein is determined by the purity, amount, and concentration of the radionuclide; the percentage of radioactivity, bound to protein with a normal biodistribution and metabolism, and the absence of microorganisms, pyrogens, particles and other substances that are toxic or might have an effect on the behaviour of the tracer proteins /16/.

Factors reviewed by Rhodes are listed in Table 10.4.

An initial problem encountered with the preparation of labelled proteins of radiopharmaceutical quality is the separation of protein-bound from unbound activity. Electrophoresis, thin-layer chromatography, paper chromatography, trichloroacetic acid precipitation have been used to determine the percentages of bound and unbound activity in the labelled protein preparation /17/. The high molecular weight artifacts, such as protein dimers or polymers, present a special problem since it is often difficult to separate them from the desired labelled protein. Analysis for high molecular weighted contaminants can be efficiently achieved on a longer (80—100 cm) column of Sephadex G-200 or Sephacryl S-200 or using high resolution capacity polyacrylamide gel electrophoresis (PAGE) technique. Application of PAGE, using reduced gels (SDS-PAGE) and non-reduced gels, provide accurate, reproducible information about the composition and activity distribution of the various labelled proteins as demonstrated in Figure 10.1. and 10.2. Retention of native biological properties and activity following the labelling and purification may be the final consideration. Biological activity can be partially destroyed due to either the conditions of the labelling procedure, the presence of the radionuclide at or near the active site of the protein or radiolysis. The most efficacious test for a labelled protein preparation is determined by its *in vivo* behaviour. Figures 10.3. and 10.4. demonstrate the different blood clearance of various protein compounds.

Table 10.4. Control data for radiolabelled proteins (ref. /16/)

A. Radionuclides
   1. Purity
   2. Quantity
   3. Concentration

B. Radiochemicals
   1. Percentage insoluble radioactivity
   2. Percentage unbound radioactivity
   3. Labelling yield
   4. Specific activity
   5. Biodistribution

C. Radiopharmaceuticals
   1. Sterility
   2. Pyrogens
   3. Particles
   4. Toxins

3. Summary and conclusions

Labelled protein derivatives used in nuclear medicine or biological research have a great potential for a beneficial application. Recently, labelled monoclonal antibodies (MOAB) as a new class of protein radiopharmaceuticals

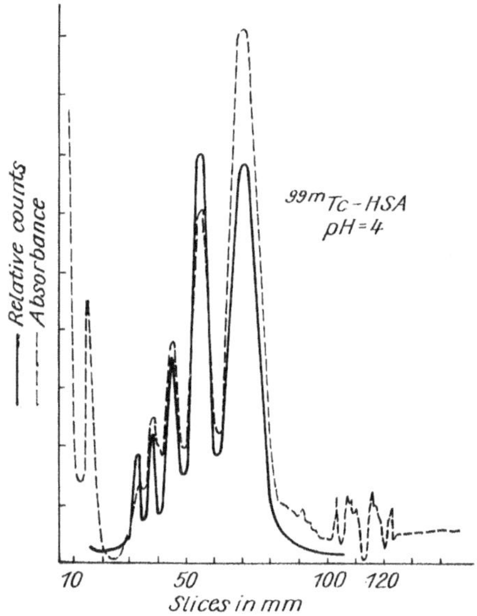

Fig. 10.1. Composition and activity distribution of ^{99m}Tc-HSA in polyacrylamide gel. Activity recovered in protein bound: monomer 37 %, dimer 27 %, trimer 13 %, polymers abt. 20 %

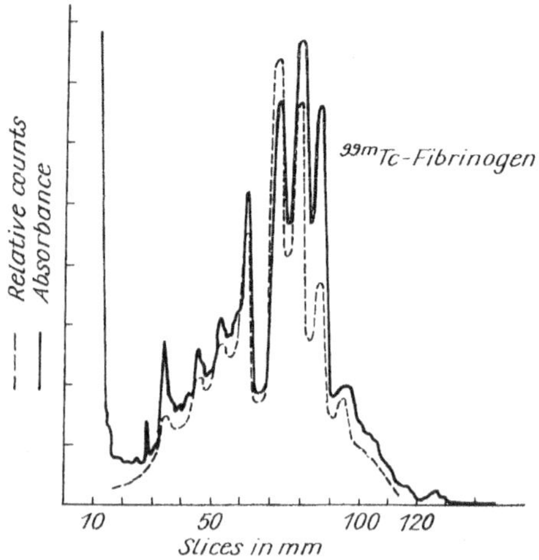

Fig. 10.2. Composition and activity distribution of ^{99m}Tc-Fibrinogen in SDS polycrylamide gel. Activity recovered in A α, B β and γ-chains about 60 %

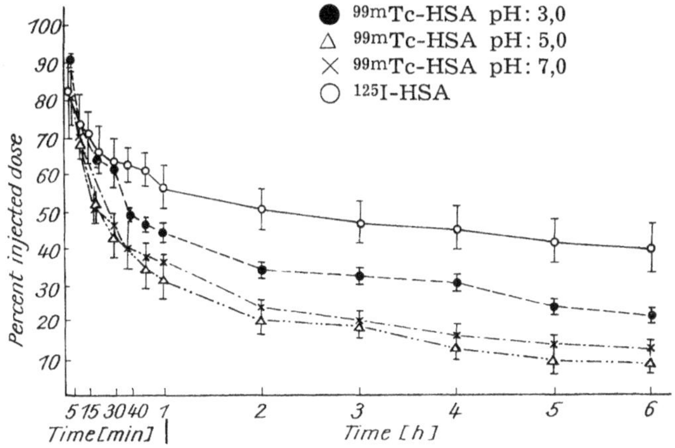

Fig. 10.3. Blood-clearance curves of different labelled HSA preparations tested in mongrel dogs using standard blood-sampling methods

have been developed. The development of monoclonal antibodies, which recognize tumour-associated antigens, has led to significantly greater practical possibilities for producing highly specific radiolabelled antibodies for diagnosis and therapy of human tumours.

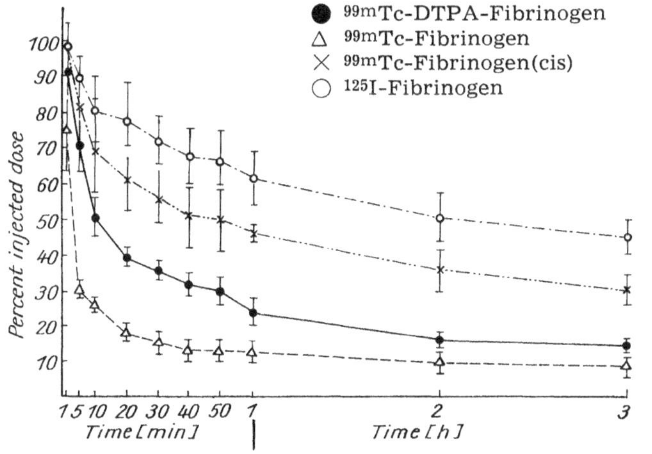

Fig. 10.4. Blood-clearance curves of different labelled fibrinogen preparations in New Zealand white rabbits using standard blood-sampling methods (Δ ^{99m}Tc-Fibrinogen tested after expiry)

References

1. *McFarlane, A. S.:* Efficient trace-labelling of protein with iodine. Nature, **182** (1958), 53.
2. *Hunter, W. M., F. C. Greenwood:* Preparation of Iodine-131 labelled human growth hormone of high specific activity. Nature, **194** (1962), 495.
3. *Redshaw, M. R., S. S. Lynch:* An improved method for the preparation of iodinated antigens for radioimmunoassay. J. Endocrinol., **60** (1974), 527.
4. *Pennisi, F., U. Rosa:* Preparation of radioiodinated insulin by constant current electrolysis. J. Nucl. Biol. Med. **13** (1969), 64.
5. *Marchalonis, J. J.:* An enzymatic method for the trace iodination of immunoglobulins and other proteins. Biochem. J. **113** (1969), 299.
6. *Fraker, P. S., K. I. Speck:* Protein and Cell Membrane Iodinations with a sparingly soluble chloramide 1, 3, 4, 6-Tetrachloro-3α,6α-diphenylglycouril. Biochem. Biophys. Res. Commun. **80** (1978), 849.
7. Iodo-Beads^{TM} Pierce Chemical Company Bio-Research Products. Technical Bulletin, 1983.
8. *Bolton, A. E., W. M. Hunter:* Labelling of proteins to high specific radioactivities by conjugation for a I-125-containing acylating agent. Biochem. J. **133** (1973), 529.
9. *Williams, M. S., T. Deegan:* The process involved in the binding of technetium-99m to human serum albumin. Int. J. Appl. Radiat. Isot. **22** (1971), 767.
10. *Persson, R. B. R., K. Liden:* Tc-99m-labelled human serum albumin. A study of the labelling procedure. Int. J. Appl. Radiat. Isot. **20** (1969), 241.
11. *Lin, M. S., H. W. Winchell, B. A. Shipley:* Use of Fe(II) or Sn(II) alone for technetium labelling of albumin. J. Nucl. Med. **12** (1971), 204.
12. *Benjamin, P. P.:* A rapid and efficient method of preparing Tc-99m-human serum albumin: Its clinical applications. Int. J. Appl. Radiat. Isot. **20** (1968), 187.
13. *Echelman, W. C., G. Meinken, P. Richards:* Tc-99m-human serum albumin. J. Nucl. Med. **12** (1971), 707.
14. *Sundberg, M. W., Ph. D. Dissertation,* Stanford University, 1973.
15. *Yoh, S. M., D. G. Sherman, C. F. Meares:* A new route to "Bifunctional" Chelating agents: Conversion of aminoacids to analogs of ethylenedinitrilotetraacetic acid. Anal. Biochem. **100** (1979), 152.
16. *Rhodes, A.:* Considerations in the radiolabeling of Albumin. Seminars in Nucl. Med. **4** (1974), 281.
17. *Pettit, W. A., F. H. Deland, L. Blanton:* Labeled proteins: Criteria for quality. Radiopharmaceuticals II. The Soc. of Nucl. Med. Inc. New York (1979), p. 33.

Chapter 11

Labelling of Proteins with Iodine-125 as RIA-indicators

H. Eberhard Schmidt, Guenter K. Vormum

Tracers in a narrow sense are used in RIA in the form of labelled ligands, in IRMA, however, as labelled antibodies (binders) for the quantification of both the bound and the free fraction of the binder or ligand. This presupposes that the binding behaviour between ligand and binder will be only insignificantly influenced by the radioactive label. In this connection we differentiate between "internal" and "external" labelling. As far as internal labelling is concerned, only an inactive atom is replaced by a radioactive isotope in the molecule of the ligand, such as in triiodine thyronine or thyroxine iodine-127 by iodine-125.

"Internally" labelled tracers biochemically and immunochemically react with the non-labelled ligand completely analogously. In this connection we can exclude any isotopy effect. Iodine is usually not a structural component of native proteins of the human organism, except for intermediate products of the thyroid gland. The introduction of iodine as a "foreign atom" raises the question whether this "external" labelling does not change the biochemical properties in an inadmissible way, or whether the labelling procedure itself produces a considerable change in the protein structure and, thus, also in the biological activity.

The increased use of radioiodine-labelled tracers in RIA kits is mainly based upon practical considerations, which result from the nuclearphysical and chemical properties of radioiodine isotopes. Because the physiological concentrations of biologically active substances are often extremely low, their precise radioimmunological determination requires tracers with high specific radioactivity. This presupposes that the radionuclide to be used for labelling is also made available with very high specific activity. Of the two most important radioiodine isotopes, iodine-125, which can be produced almost carrier-free, has prevailed as the radionuclide preferred for labelling (Table 11.1.). Also its favourable physical half-life of 60 days, which makes it possible to use a ^{125}I-labelled tracer for several weeks, as well as its radiation properties (no corpuscular radiation) favour it compared to the other radioactive iodine isotopes. In addition to this, the radiation risk occurring during work with the soft gamma-emitter iodine-125 is lower, despite the threat of incorporating the longer-lived radionuclide.

Another reason for the increased application of radioiodine isotopes in protein labelling is due to their chemical reactivity. There are a number of preparative techniques for the synthesis of organic iodine compounds, which in a proper modification allow the introduction of radioiodine into albumins chemically or enzymatically under particularly mild and preserving conditions and by largely maintaining immunological reactivity. Moreover, if

Table 11.1. Properties of $Na^{125}I$ and $Na^{131}I$

	$Na^{125}I$	$Na^{131}I$
Physical half-life	60.0 d	8.05 d
Maximum specific activity at 100 % isotope abundance	644 GBq/mg	4,580 GBq/mg
Specific activity of commercial preparations	> 555 GBq/mg	150–500 GBq/mg
Mode of nuclear radiation and energy	EC γ: 0.085 MeV	β^-: 0,61 MeV γ: 0,36/0,64 MeV
Counting efficiency in the NaI borehole scintillation counter	50–70 %	30–40 %

we consider that also the separation of the reaction products and their analytical purity control does not in general involve any difficulties, it becomes clear that iodine-125 is in fact a radionuclide with highly favourable properties.

Depending upon the structure of the antigen to be labelled, there are different possibilities of incorporating a radioiodine atom into a protein molecule during a chemical reaction:

- Higher molecular proteins contain amino acids as primary structural components, in which an electrophilic iodine substitution can be performed directly, i.e. without any modification of the initial substance.
- Before labelling, a residue, which is able to be iodinated, is fixed at and/or incorporated into the polypeptides, which do not have any suitable position for iodination.
- In a third variant, the radioiodine atom is first bound to a carrier molecule which is immunologically not related to the antigen molecule, but it possesses reactive groups so that it can be coupled to the protein molecule in a second reaction step.

Thus, we distinguish between direct and indirect labelling techniques, whose advantages and disadvantages will bei discussed in the following.

The mainly used chemical reaction occurring in direct radioiodination of a protein is an electrophilic substitution reaction on the tyrosine residues of the amino acid chain (Figure 11.1.). The iodine atom takes the o-position towards the hydroxyl group of the aromatic ring. Radioiodine is oxidized in a buffer solution in the presence of the protein to be labelled. This formulation is to describe the ionic mechanism between the iodine cation formed and the phenolic anion. Depending on the stoichiometric ratio of the radioiodine to the protein quantity, the reaction mixture contains, apart from the monoiodine substitution product, also diiodinated tyrosine residues.

We can assume that the incorporation of an average of 1 iodine atom per protein molecule produces a negligible substitution effect on the immunological properties of an antigen. Thus, a radioiodine labelling should be carried out in such a way that mainly monoiodine substitution products are formed. This in turn provides a criterion for the optimum specific activity.

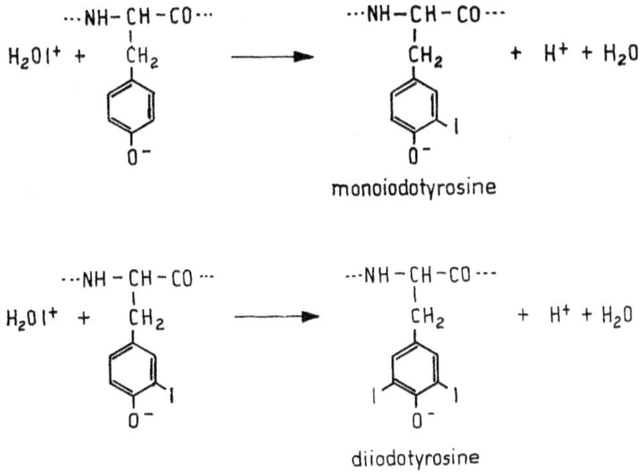

Fig. 11.1. Structure formula of Tyrosine

Table 11.2. shows the relevant values for several peptide hormones and glycoproteins. However, we have to take into account that the yet admissable degree of substitution also depends on how much tyrosine residues a protein molecule contains, how these are distributed in the amino acid chain, and how reactive they are. It has been found that tyrosines, which are easily accessible, are iodinated preferably. If these "external" tyrosine residues happen to be located on those positions of the peptide chain, which are involved in the formation of the antigen-antibody-complex, an iodine substitution at this point of the molecule will certainly result in a reduction of the antigen effect. The probability that the radioiodination involves such a "critical" tyrosine residue will grow depending on the increase in the average degree of substitution.

Table 11.2. Specific activities of ^{125}I-labelled hormones of different molecular weight after radioiodination with carrier-free Na^{125}I up to an average degree of substitution of 1 radioiodine atom per molecule

Hormone	mol. weight	Tyr/ molecule	MBq/μg
LH releasing hormone	1,182	1	61.6
Gastrin (human)	2,116	1	35.9
Glucagon	3,485	2	22.3
Human corticotrophin (ACTH)	4,541	2	17.3
Insulin (bovine)	5,780	4	13.6
Human chorionic gonadotrophin (HCG)	21,000	8	3.80
Human chorionic gonadotrophin (HCG)	30,000	8	2.67
Human luteinizing hormone (HLH)	33,000	8	2.43

Apart from this reaction occurring predominantly in neutral or weak-alkaline media, there are a few competitive reactions of the oxidized radioiodine to other amino acid residues of the protein as well as to functional groups which can be easily oxidized (Figure 11.2.). Thus, also an electrophilic exchange of hydrogen for iodine on the imidazole ring of the amino acid histidine as well as on the indole ring system of tryptophane can occur. The reaction velocity of histidine iodination is, however, considerably lower than that of tyrosine iodination; the difference is even greater compared to tryptophane. When investigating radioiodination of tyramine-histamine-mixtures as model substances we have found that the relative reaction velocity of tyramine iodination depends upon the pH-value. Tyramine reacts almost 100 times faster than histamine (Figure 11.3.).
Among the easily oxidizable functional groups of a protein molecule, the nucleophilic SH-residues will always react first under the effect of oxidized iodine (Figure 11.4.). In the case of a peptide molecule this can result in the formation of disulphide. Considering a higher molecular protein, which for steric reasons cannot form a disulphide chain, the sulphenyle iodide formed hydrolyzes rapidly. This somewhat schematized representation shows that always only a portion of some 60 up to 80 per cent of the radioiodine applied is bound to the protein. The share in iodine at first used for non-specific oxidation of the SH-groups must be regarded as a loss for the actual labelling reaction. The quantity of oxidizing agent required cannot therefore be rated stoichiometrically but it must be so large as to ensure before the main reaction that both the low portions of reducing agents, which can often be found in radioiodine solutions, are neutralized and the side reactions, which are partially unavoidable and occur more rapidly, are taken into account. This effect is shown in Figure 11.5. The iodination of 100 μg bovine serum

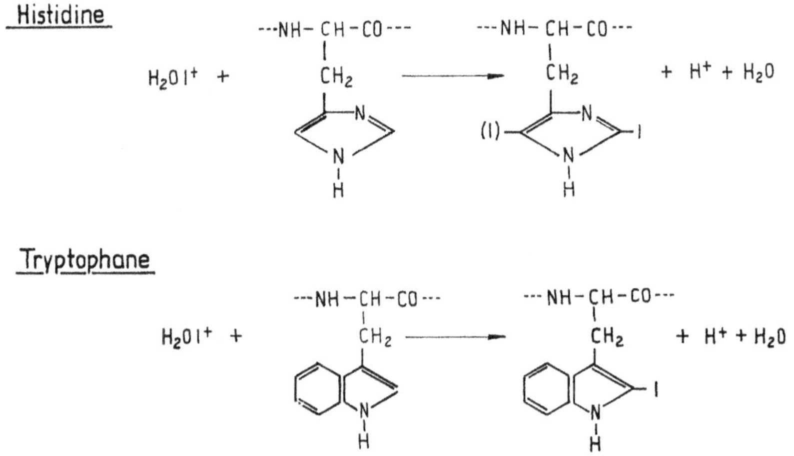

Fig. 11.2. Structure formula of Histidine and Tryptophane

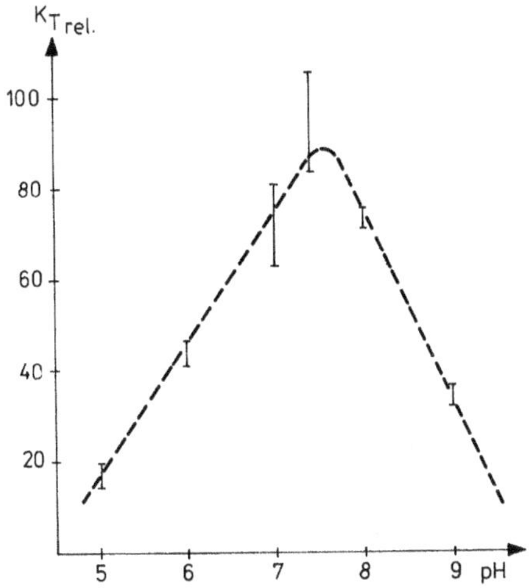

Fig. 11.3. Relative reaction velocity ($K_{T\ rel.}$) of tyramine iodination determined by means of paper chromatographic separation of radioiodinated tyramine-histamine-mixtures in dependence upon the pH-value

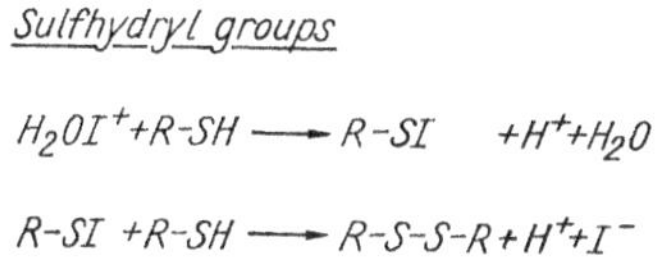

Fig. 11.4. Sulfhydryl groups

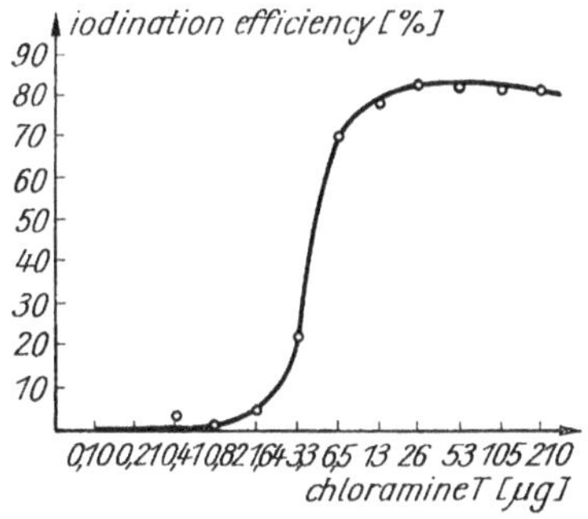

Fig. 11.5. Ineffectiveness of radio-iodination of 100 μg BSA with too low a chloramine T-concentration. Taken from: Weir, D. M. (Ed.): Handbook of Experimental Immunology. Blackwell Scientific Public., Oxford – Edinburgh 1973.

albumin in each case at low chloramine T-concentrations does not produce any effective incorporation of iodine. Only when using at least 5 μg chloramine T, the yield curve increases drastically and then reaches an even plane. Even if oxidants are excessively used, a higher efficiency of iodination cannot be achieved.

Over the past 25 years labelling methods have been continuously improved in order to be able to manufacture radioiodine-labelled tracers with high specific activity for RIA-use while largely retaining their immunological properties. Table 11.3. provides a survey of the most important techniques for direct or indirect radioiodination. Moreover, it lists the research groups which have developed them. Considering the direct labelling methods, it should be emphasized that they differ by the oxidant applied.

In 1958 McFarlane /1/ developed the first practical method for the radioiodination of proteins. He used iodine monochloride which is largely ionized in diluted hydrochloric and/or chlorideionic solution, thus being more reactive than elementary iodine (Figure 11.6.). It is prepared in situ from potassium iodide and potassium iodate in hydrochloric solution. It is a disadvan-

Table 11.3. Technique of radioiodine labelling of proteins and polypeptides

Oxidizing agents for direct radioiodination	
1. Iodine monochloride	McFarlane (1958)
2. Elementary iodine	Yalow and Berson (1960) Banerjee and Ekins (1961)
3. Chloramine T	Hunter and Greenwood (1962)
4. Electrolytic iodination	Rosa et al. (1964)
5. Enzymatic iodination using	
– horseradish peroxidase	Nunez et al. (1965)
– lactoperoxidase	Marchalonis (1969)
6. Chlorine gas	Butt (1972)
7. Sodium hypochlorite	Redshaw and Lynch (1974)
8. 1,3,4,6-Tetrachloro-3α,6α-diphenyl-glycoluril (= Iodogen)	Fraker and Speck (1978) Salacinski, Hope and McLean (1979)
Indirect radioiodination by conjugation with radioiodine labelled reagents	
1. 3-(p-Hydroxyphenyl)-propionic acid-N-hydroxysuccinimidyl ester (= Bolton-Hunter-reagent; Tagit)	Bolton and Hunter (1972)
2. Methyl-p-hydroxybenzimidate hydrochloride	Wood, Wu and Gerhart (1975)
3. Diazotized aniline	Hayes and Goldstein (1975)

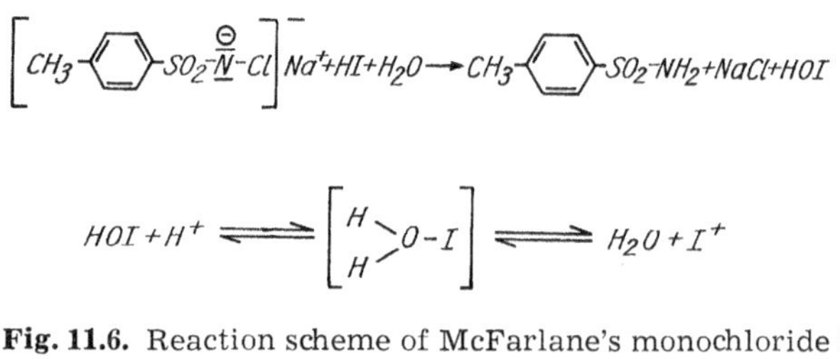

Fig. 11.6. Reaction scheme of McFarlane's monochloride iodination method /1/.

tage of this method that with potassium iodate also inactive iodine is involved in radioiodination. Thus, no high specific activit cany be achieved.
In contrast to this, despite several objections raised time and again in the literature, direct protein iodination with chloramine T, which was developed by Hunter and Greenwood /2, 3/ in 1962, has been widely accepted. This is mainly due to its fastness and the limited technological equipment necessary for using this labelling method. Chloramine T is the sodium salt of the N-monochloro-p-toluenesulphonamide. It acts as a mild oxidizing agent leading to a highly effective organic binding of radioiodine even at very low concentration. The mechanism of this reaction is not known exactly. The formation of iodine cations probably occurs via the intermediate stage of the hypoiodite (Figure 11.7.).

$$2KI + KIO_3 + 6HCl \longrightarrow 3ICl + 3KCl + 3H_2O$$

Fig. 11.7. Reaction scheme of chloramine T-iodination method according to Hunter and Greenwood /2, 3/

Because the efficiency of iodination depends largely on protein concentration, the reaction volume is kept as small as possible considering the small protein quantity, e.g. 5 μg, in any case below 100 μl. In addition to this, the labelling yield depends upon the pH-value of the reaction mixture. Most proteins have a pH-value-optimum of between 7 and 7.5. The efficiency of incorporation of iodine is also influenced by the concentration and duration of the effect of chloramine T. In order to avoid a non-specific damage to the protein, both the reaction parameters should be kept at a minimum. The reaction times cover between 10 and 60 seconds. The more sensitive a protein to oxidation is, the faster radioiodination will be stopped by adding a reducing agent. Usually sodium disulphide is still being used for this purpose, and this with a marked excess compared to the quantity of chloramine T used. It is obviously assumed that bisulphite ions in low concentration do not exercise any damaging effect. Past experience has shown however that proteins, in particular antibodies, can be otherwise adversely affected in their immunological reactivity even at relatively low sulphite concentrations connected with the reduction of disulphide bindings through partial sulphite lysis of the molecule. In this connection it should be pointed out that these disulphide bindings are responsible for the tertiary structure of the molecules. For this reason, since 1974 we have used 2-mercaptoethanol for stopping; there are numerous data in the literature which report on its milder reduction effect vis-a-vis antibodies. However, the amount of 2-

mercaptoethanol should be kept as small as possible; a 1.5- up to 2-fold stoichiometric excess compared to chloramine T will suffice. The optimum labelling conditions will have to be found for each protein through preliminary tests.

Enzymatic radioiodine labelling, which was developed by a research group under Nunez and colleagues /4/, is an interesting and highly preserving method. With this method radioiodide is oxidized by traces of hydrogen peroxide. At the beginning horse-raddish peroxidase was used as a catalyst. In 1969 Marchalonis /5/ succeeded in improving the technique substantially by using the enzyme lactoperioxidase.

As far as enzymatic radioiodination is concerned, the protein, i.e. about 5 μg, is mixed with $Na^{125}I$ and 50 up to 100 ng of lactoperoxidase in a weakly acid buffer solution. Just as with iodination with chloramine T, we work with the smallest possible reaction volume. By adding 50 up to 100 ng hydrogen peroxide to a fresh-diluted solution, radioiodination is initiated and continued by one or two further H_2O_2 additions of the same quantity, which are added at an interval of about 2 up to 5 minutes. The reaction can be stopped by adding cysteine or sodium acide which eliminates the catalytic effect of lactoperoxidase, or simply by dilution. To understand the enzyme effect, one has to consider that certain surface regions will get into direct contact with the protein to be labelled. This sterically oriented reaction markedly increases the probability of obtaining in particular monoiodinated tyrosine residues.

As an alternative to the chemical radioiodination techniques available, Rosa and colleagues /6/ have introduced electrolytic iodination which is a very interesting variant in terms of technical procedure. The iodine cations are generated by a controlled anodic oxidation of the radioiodide solution, which is to be found together with the protein solution to be labelled in a microelectrophoresis cell made from platinum. Neither oxidizing nor reducing agents are used here. However, the relatively long time of electrolysis, which can lead to the denaturation of sensitive proteins, is adverse.

New is the concept to carry out radioiodination as a solid-phase reaction so as to prevent the protein to be labelled from getting into direct contact with the oxidation reagent. From this point of view, in 1978 Fraker and Speck /7/ for the first time reported on protein iodinations by using chloramide 1,3,4,6-tetrachloro-3*a*-6*a*-diphenylglycouril (Figure 11.8.), which is practically insoluble in an aqueous medium. In 1981 the American producer Pierce Chemical Company put it on the market under the trade-name IODOGEN /8/. Meanwhile quite a few articles have been published which report on the partially excellent results achieved by this new labelling technique /9—17/.

The implementation of solid-phase radioiodination is very simple. Iodogen is put into a labelling tube of glass as a solution in an organic solvent. After the solvent has evaporated, it adheres as a stabile and firm film to the bottom and/or the inner side of the tube. The reaction tubes covered with such a film can be stored for a long time. The protein to be labelled is in a buffer solution; then it is pipetted into the tube and after adding radioiodide solution it is allowed to settle at room temperature for a short time (1 up to

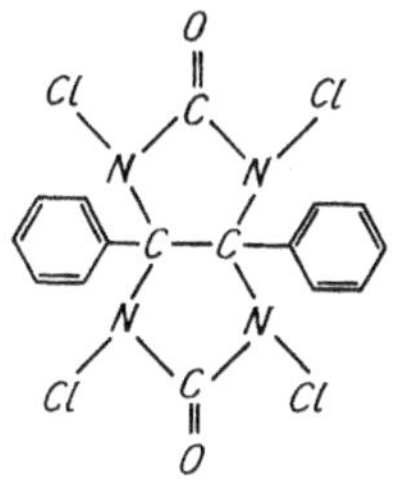

Fig. 11.8. Structure formula of Iodogen (TM = Trade mark) (Fraker and Speck 1978) Pierce Chemical Company, Rockford (USA): Iodogen TM

5 min.), while occasionally shaking the tube gently. Then the labelling mixture is simply removed with a pipette; in this way the reaction is stopped. In association with the immobilization of the oxidizing agent, which establishes direct contact with the protein, it is stated as an important advantage of this technique that no reducing agents are used. Direct contact with the protein can produce undesired side reactions, which in turn might lead to a structural change including a loss in immunoreactivity.

Indirect radioiodination does not involve those problems, which may occur in direct labelling procedures through oxidizing or reducing agents as well as through iodine substitution on the tyrosine residues which are critical for antigenity. It has already been mentioned that at first a proper carrier molecule is radioiodinated and after purification it is coupled to the protein molecule. The position for the radioactive partner, which possesses a reactive group, are the ε-amino groups of lysine or the N-terminal NH_2-group of the protein molecule. Thus, in this way also peptides, which do not possess any tyrosine residues for direct iodination, can be labelled. Such a coupling reaction, however, largely depends on the concentration concerned. Consequently, even if the smallest possible reaction volumes are used, only low yields can be obtained. It is only useful to prepare radioiodine tracers of high specific activity by a conjugation method, if one can use a larger quantity of substance. Moreover, if one takes into consideration that the practical implementation of such a two-step reaction requires more time and is more complicated, because more manipulations with radioactive material are involved, it becomes obvious why the radioiodination of proteins for RIA-applications based upon a coupling method has been confined to special cases so far despite the proven immunoreactivity of the preparations. Thus, as far as indirect radioiodination techniques are concerned, only the method with the reagent developed by Bolton and Hunter /18/ in 1972, which is named after them and goes under the trade-name TAGIT, will be discussed here.

Tagit possesses both a p-hydroxyphenyl group, which can easily be subject to electrophilic radioiodine substitution as with tyrosine residues, and an active ester group providing reactivity with nucleophilic NH_2-groups (Figure 11.9.). As Tagit hydrolizes very rapidly in aqueous solution, the initial radioiodination with chloramine T is carried out as quick as possible in order to keep the hydrolysis occurring under the formation of p-hydroxyphenylpropionic acid down at an extreme minimum. The ^{125}I-HPNS has to be extracted from the aqueous phase with benzene without any delay of time, separated simultaneously from non-bound radioiodide as well as by-products of chloramine T-iodination, and then dried. The second step of reaction, i.e. the coupling to the protein, occurs in weak-alkaline solution. In order to push back hydrolysis vis-a-vis aminolysis, an ice bath should be used. After a reaction time of up to 3 hours, glycine is added with a large excess so as bind the ester which has not yet been transformed.

Summarizing we state that quite a number of techniques have been developed for radioiodine labelling of protein tracers. All these methods have advantages and disadvantages; an optimum labelling method providing entirely satisfactory results for each protein concerned at the first attempt does not exist.

Regardless of the labelling method chosen it is necessary to separate the radioiodinated protein from radiochemical by-products, which either are contained in inactive form even in initial material and which can be detected only after labelling or which are formed as so-called degradation products during radioiodination. Furthermore, the radioiodinated protein needs to be separated from non-bound radioiodide and also from the labelling reagents applied. For this purpose, reliable and widely accepted preparative methods of protein chemistry are used. In most cases the reaction mixture will be subject to gel filtration by making use of the molecular sieving properties of dextrane gels. However, also other chromatographic techniques, such as thin-layer or column chromatography, are used. Moreover, ion-exchange chromatography, the chemical precipitation method or the electrophoretic separation technique are applied.

In conclusion, we want to briefly discuss the question connected with the stability of radioiodinated proteins and their limited time of applicability. Even at storage in deep-frozen state, a decrease in immunoreactivity cannot be avoided in the course of 4 up to 5 weeks. Radiochemical purity checks of the tracers stored over this period of time have found that besides a loss in immunological reactivity the content of free iodide increases continuously. Attempts to reduce this time-dependent loss in immunoreactivity

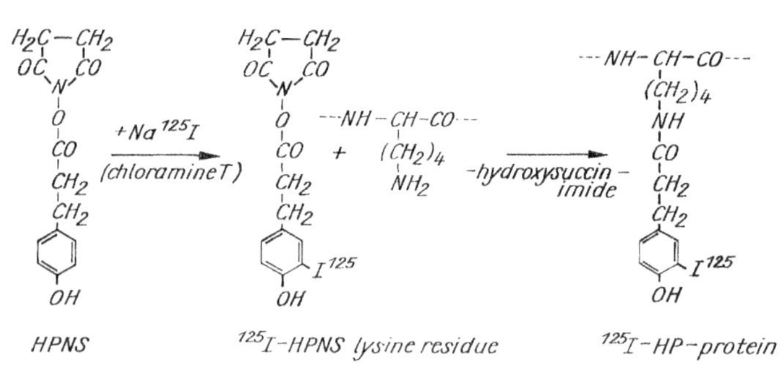

Fig. 11.9. Reaction steps in protein labelling by conjugation with ^{125}I-labelled 3-(p-hydroxyphenyl)-propionic acid-N-hydroxysuccinimidyl ester (HPNS)

by additions of various kinds have produced only partial successes. By adding suitable heterologous protective proteins as scavengers as well as by rapid freezing of small aliquotes of the tracer solution, which need to be thawed for use one time, the problem could only be minimized but not solved. Also freeze-drying, producing sometimes stabler preparations with a higher shelf-life, has not always proved to be successful.

The present state of knowledge requires to make a compromise concerning the work with higher molecular radioiodinated tracers: the specific activity should be kept within limits, radioactive concentration should not be increased too much, and a radioprotective heterologous protein, usually 0.05 per cent bovine serum albumin, should be added. If the tracer is stored in deep-frozen state immediately after preparation and also transported, it can be successfully used for RIA-applications for about 4 up to 6 weeks. Only radioiodinated peptides, which can be separated from the labelling mixture as monoiodinated compounds, can be used for a longer period of time despite their high specific activity.

References

1. *McFarlane, A. S.:* Efficient trace-labeling of proteins with iodine. Nature (London) **182** (1958), 53.
2. *Hunter, W. M., F. C. Greenwood:* Preparation of iodine-131 labelled human growth hormone of high specific activity. Nature (London) **194** (1962), 495.
3. *Greenwood, F. C., W. M. Hunter, J. S. Glover:* The preparation of I-131 labelled human growth hormone of high specific radioactivity. Biochem. J. **89** (1963), 114.
4. *Nunez, J., J. Pommier, M. E. Hilali, J. Roche:* Jodination enzymatique de proteins. J. labelled Comp. **1** (1965), 128.
5. *Marchalonis, J. J.:* An enzymic method for the trace iodination of immunoglobulins and other proteins. Biochem. J. **113** (1969), 299.
6. *Rosa, U., G. A. Scassellati, F. Pennisi, N. Riccione, P. Giagnoni, R. Giordani:* Labelling of human fibrinogen with I-131 by electrolytic iodination. Biochem. biophys. Acta **86** (1964), 519.
7. *Fraker, P. J., J. C. Speck:* Protein and cell membrane iodinations with a sparingly soluble 1,3,4,6-tetrachloro-3α,6α-dephenylglycouril. Biochem. Biophys. Res. Commun. **80** (1978), 849.
8. *Markwell, M. A. K.:* Radioiodination of proteins and peptides using iodogen. Technical Bulletin, Pierce Chemical Company, Rockford, Illinois (USA), 1981.
9. *Markwell, M. A. K., C. F. Fox:* Surface-specific iodination of membrane proteins of viruses and eucaryotic cells using 1,3,4,6-tetrachloro-3α,6α-diphenylglycouril. Biochemistry **17** (1978), 4807.
10. *Salacinski, P. R. P., J. Hope, C. McLean, V. V. Clement-Jones, J. E. C. Sykes, J. Price, P. J. Lowry:* A new simple method which allows theoretical incorporation of radioiodine into proteins and peptides without damage. J. Endocrinol. **81** (1979), 131.
11. *Tolan, D. R., J. M. Lambert, G. Boileau, T. G. Fanning, J. W. Kenny, A. Vassos, R. R. Traut:* Radioiodination of microgram quantities of ribosomal proteins from polyacrylamide gels. Anal. Biochem. **103** (1980), 101.
12. *Knight, L. C., A. Z. Budzynski, S. A. Olexa:* Comparison of the iodogen and iodine monochloride methods for radioiodination of fibrinogen. J. Labelled Comp. Radiopharm. **18** (1981), 50.

13. *Knight, L. C., A. Z. Budzynski, S. A. Olexa:* Radiolabeling of fibrinogen using the iodogen technique. Thrombos. Haemostas. **46** (1981), 593.
14. *Salacinski, P. R. P., C. McLean, J. E. C. Sykes, V. V. Clement-Jones, P. J. Lowry:* Iodination of proteins, glycoproteins and peptides using a solid-phase oxidizing agent 1,3,4,6-tetrachloro-3α,6α-diphenylglycouril (Iodogen). Anal. Biochem. **117** (1981), 136.
15. *Wood, W. G., Ch. Wachter, P. C. Scriba:* Experiences using chloramine T and 1,3,4,6-tetrachloro-3α,6α-diphenylglycouril (Iodogen) for radioiodination of materials for radioimmunoassay. J. Clin. Chem. Clin. Biochem. **19** (1981), 1051.
16. *Schmidt, H. E., W. Delzer:* Bietet die Proteinmarkierung mit Radiojod unter Einsatz von Iodogen wirklich Vorteile? Radiobiol. Radiother. **24** (1983), 866.
17. *Butler, S. R., R. W. Lam, D. A. Fisher:* Iodination of thyroliberin by use of Iodogen. Clin. Chem. **30** (1984), 547.
18. *Bolton, A. E., W. M. Hunter:* Labelling of proteins to high specific radioactivities by conjugation to a I-125-containing acylating agent. Biochem. J. **133** (1973), 529.

Chapter 12

Biological Aspects of Radiopharmacokinetics

Annemarie Blottner

Radiotracers provide a sensitive means to detect and measure a multitude of biological and chemical functions. Their applications in diagnostic medicine have introduced a number of new and effective techniques to study a myriad of pathologic conditions. Despite this progress, the most important and pressing problem in nuclear medicine has not yet been solved, i.e. the lack of specificity of a lot of radiopharmaceuticals.

A possible solution to this dilemma may be found in a better understanding of the mechanisms involved in the development of a radiotracer. To analyze the consequence of a radiotracer in a biological system, it has been found that the functional radionuclide image can provide a support for most of the reactions involved in radiotracer transport and localization.

Radiotracers may be single ions or complexes, and their individual pharmacokinetics dictate their localization and clinical use. The ideal radiotracer behaves like its unlabelled ligand, and its radiolabel does not become involved in the reaction or included into metabolites which interfere the process. Most radiotracers employed in medicine are quite dissimilar in their molecular configuration to their unlabelled ligand, which is due to radiolabel attachment.

Most radiotracers are administered directly into the blood stream. A number of mechanical and physiological factors may cause their retention or removal from the vascular space. The study of absorption, distribution, metabolism and excretion, dependent on the time, is called pharmacokinetic.

The major problem in pharmacokinetics is the question about the relationship between the blood levels and the pharmacological effect of drugs. Therefore, the main question is: How can the change in concentration be determined, calculated and explained? As a rule, drug concentration in the study of unlabelled pharmaceuticals for therapeutical treatment is determined in parts of the organism, which are easily accessible, e.g. in blood or excrements. These measured data are approximated through model curves by means of kinetic models. In this way aim to predict the courses of concentration in compartments, which are difficult of access or not accessible for direct measurements. The application of radioactively labelled compounds makes it, on the other hand, possible to measure radioactivity in most parts of the organism. The measured data provide the foundation for possible functional tests in healthy and damaged organs. It is also advantageous that the possibility of application of least quantities of radioactively labelled drugs—the level lies on the physiological range or much lower—allows to watch the kinetical processes in the organism without any change in the functions of organs.

Before the clinical application of radiopharmaceuticals one has to analyze, whether the metabolic processes result in a separation of the radioactive isotope from the unlabelled ligand. Considering these processes, the measurement of the emerging metabolic products, which depend on the period of time, can lead to important conclusions on the function of organs and organ systems. The same fact is valid for the application of labelled precursors, which requires to know the synthesis process.
Successful research in this field requires the knowledge and control of a great number of methods, which are not available or are only limited at the departments of nuclear medicine. The emphasis of radiochemical and radiopharmacokinetical examination at those departments is often placed therefore on the production and the experimental comparison of radiopharmaceuticals, whose usefulness in clinical application was accepted by other internationally acknowledged working groups.
We confined ourselves to working out data tables and the verbal description of kinetic progressions. The main questions of these examinations were:

1. How quickly does the radiopharmaceutical leave the blood stream, applied by i.v. injection?
2. How high is the uptake of radiopharmaceuticals in different organs?
3. How long is the retention of radiopharmaceuticals in organs and organ systems?
4. Is a metabolic variation of radiopharmaceuticals after organ passage demonstrable?

Recently there has been much discussion about the necessity of animal experiments. But there is no real substitute for animal models available because their level of organization is more complicated than that in tissue cultures or cell suspensions.
We mainly used for experimental work rats and mice as laboratory animals, because these studies demand a high quantity. For pharmacokinetical examinations out-breeding animals are used. These animals have a natural but constant degree of variableness of their manner of reaction. The degree of heterozygoty is preserved. The keeping of animals requires a definite environment, because all organ functions must not be injured. Some important parameters are: climate of room and cage, temperature, acclimatization after transport, air moisture and noise. Before we use the animals in experiments, they have to be kept under constant conditions for at least 7 days.
As the pharmacokinetical parameters are determined not only by radioactive drugs, but also by the biological organism, differences in species can be expected, and we have to be cautious about the transfer of the experimental data, obtained in animals, to the human organism.
Table 12.1. to 12.3. show some important physiological data in laboratory animals in comparison to the human organism. The frequency of respiration and the heart rate are in mice higher by the factor 10.
The blood volume, referred to 100 g of tissue, does not vary strongly in different species.
Only small distinctions can be observed in blood perfusion of the kidney and liver, but a great difference for the brain.

Table 12.1. Some important physiological data in different species

Species	Frequency of respiration	Heart rate/ minute	Temperature	Body weight
Human	10– 15	60– 80	36.2–37.4	50– 70 kg
Rat	110–150	200–370	32.1–38.1	150–300 g
Mouse	140–210	520–780	37.9–39.2	20– 26 g

Table 12.2. Some important physiological data in different species

Species	Plasma volume ml/100 g	Cell volume ml/100 g	Blood volume ml/100 g
Human	4.5	2.5–3	7
Rat	3.5	2	5.5
Mouse	5.5–6.3	2.9	6.6–7.6
Rabbit	3 –5	1.7–2	5.5–7
Dog	5.5	~ 3.8	9

Table 12.3. Some important physiological data in different species

	ml blood/organ min		% HMV	
Organ	human	rat	human	rat
Kidney	1,200	6.9	22	17.8
Liver/bowels	1,800	12.3	33	25.2
Brain	750	0.7	14	1.2

The studies of the distribution of radiopharmaceuticals serve the purpose to determine the kinetical behaviour of the substance in the organism. After i.v. injection the compounds are first distributed in the space of the vessels. Then they are transferred to the other spaces of distribution. In Figure 12.1. the possible compartments and their percentages of the whole body, in which the drug can be distributed, are shown.

The mechanism for radiotracer uptake and retention in these different compartments includes, among other things, the similarity of the radioligand to a circulating ligand, the increased molecular size and the degree of intravascular protein binding.

The concept that serum protein binding of the radiotracer can promote its intravascular retention has been used to explain a number of observations. *The ionic composition of the serum proteins* (OH^- and COO^- groups), *their internal molecular structure* (the distance between hydroxyl and carboxyl groups) and *the environment of the system* (pH, protein and anionic concentration, temperature) in which the radiotracer is placed, determine the degree of protein binding. Serum albumin is the most avid binder, but also α- and β-globulines are able to bind endogenous and exogenous compounds.

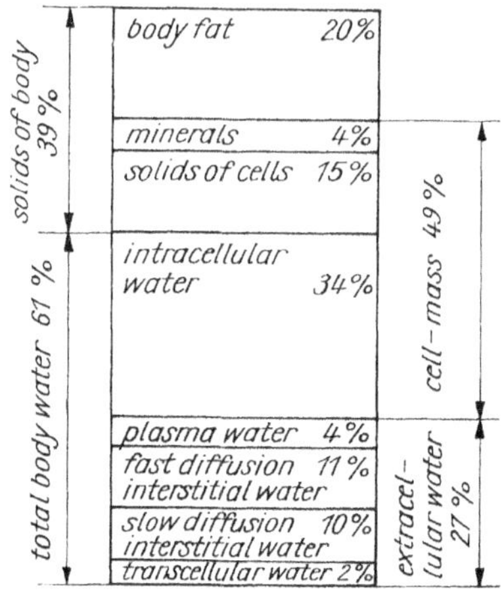

Fig. 12.1. Shares of individual distribution phases in whole-body quantity

The plasma proteins are able not only to adsorb basic but also acid and neutrally reacting substances respectively. For each compound a specific equilibrium between the free and bound portion occurs. Lipid-soluble substances are preferrably transported by the lipid fraction of the blood or by erythrocytes.

Within a class of chemical compounds, protein binding is used for the characterization of derivates, because this can be *one* reason for differences in distribution. Normally the penetration of radioactive drugs from blood into tissues takes place very quickly. The equilibration time of penetrable substances very often amounts to about 10 minutes. It *can* be delayed by a high protein binding. Apart from this, a difference in binding affinity can exist, although the degree of protein binding is the same.

The data for blood level curves and in vitro protein-binding are obtained by the procedure performed in the following manner. After i.v. injection of the radiopharmaceutical into the tail vein, we again tap 15 seconds p.i. another tail vein, and take away 10 μl blood. Further blood *drawings* are performed after 30 and 60 seconds, 5, 10, 15, 30, 45 and 60 minutes, 2, 4, 6 and 24 hours. In most cases the graphic analysis of the curve shows that radiopharmaceuticals are distributed in several compartments. If the measured data are plotted semi-logarithmically, we obtain a curve, which consists of two or more exponentially proceeding processes (α- and β-phase), representing the distribution and elimination as connected occurrences.

$$\beta = \frac{\ln c_{p1} - \ln c_{p2}}{t_2 - t_1}$$

(elimination rate constant)

$$\alpha = \frac{\ln c_{p1}' - \ln c_{p2}'}{t_2' - t_1'}$$

(distribution rate constant)

Figure 12.2. shows as an example the blood elimination of Tc-99m-diethyl-IDA; the values of radioactivity were obtained between 30 seconds and 6 hours. We received after graphic analysis *3 part functions* with different half-lives of elimination.
For Tc-99m-diethyl-IDA we calculated the value of 0.149 (h^{-1}) for the elimination constant within the time of observation.
The in vitro binding to plasma proteins was determined by equilibrium dialysis. Another possibility would be the column chromatographic analysis. We used home-made dialysis chambers. They are made of piacryl and have a capacity of 1,000 μl. The radiopharmaceutical, in phosphate-buffered physiologic saline solution, is separated from the protein through a cellophane membrane. They are cleaned before used and kept in distilled water and than put into the buffer solution to be used. The mixing can be carried out through shaking or rotation at a constant temperature. Firstly, we have to estimate the final point of dialysis in a protein-free system, i.e. the time, when there is in both parts of the chambers the same radioactivity. From this examination we can calculate the unspecific binding. Simultaneously, the in vitro stability of the preparation is tested under the same conditions by electrophoretical or chromatographic methods. After that the degree of protein binding, for example to human serum albumin (HSA), human serum gamma globulin or human serum, is determined.
Table 12.4. demonstrates the influence of the pH of the preparation on protein binding in the case of Tc-99m-thioglucose. Preparations labelled between pH 4 and 7 show about the same maximal uptake in the kidney, the target organ; the value varies between 25 and 30 % of dose, but at pH 3 and 9 we could find only 10 % in the kidney.

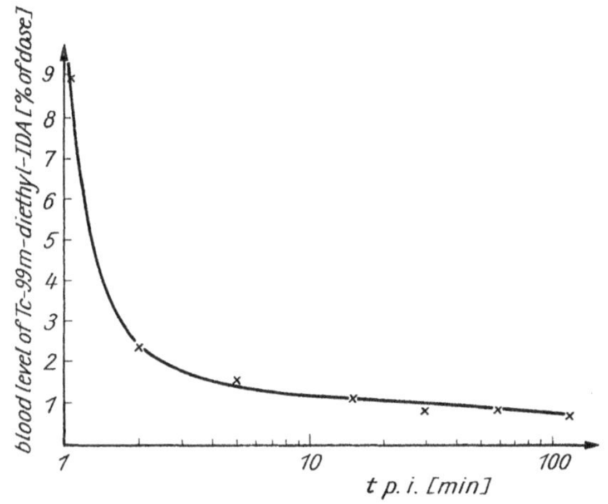

Fig. 12.2. Blood elimination (cpm/ml) in ICR-mice

Table 12.4. In vitro binding of Tc-99m-thioglucose to plasma proteins, phosphate-NaCl-buffer, pH 7.4 at 20 °C, equilibrium dialysis

pH of preparation	% binding to proteins
3	79
4	75
7	65
9	32

Table 12.5. In vitro binding of Tc-99m-diethyl-IDA to different proteins, phosphate-NaCl-buffer pH 7.4, 20 °C, equilibrium dialysis

Protein	% binding
1 % HSA	40
1 % HGG	22
human serum (1 : 7 diluted)	34

Table 12.6. In vitro protein-binding of Tc-99m-IDA-derivates phosphate-NaCl-buffer, pH 7.4, 20 °C

Protein	Tc-99m-diethyl-IDA	Tc-99m-2,4,6-trimethyl-IDA	Tc-99m-4-brom-IDA	Tc-99m-methyl-IDA
HSA 4 %	45 %	23 %	65 %	18 %
Serum	35 %	14 %	60 %	10 %

Another example are some compounds, which were studied in comparison to Tc-99m-diethyl-IDA in connection with the optimization of this radiopharmaceutical, used for the diagnosis of the bile duct. Beside the results, you can see in Table 12.5. and 12.6., we could determine the same level of protein binding in a HSA-HGG-solution, mixed in a proportion of 60 : 40.
On the basis of these findings we tried to find a relationship between serum protein binding and organ distribution. On *quite different* compounds with distinct protein binding the results do not demonstrate a dependence between protein binding and uptake in the organs. Three of the tested substances have about the same liver uptake and urinary excretion rate (Table 12.7.).
It can be concluded that the binding of radioactive drugs to tissue proteins can be *more important* then to plasma proteins. This binding also influences the pharmacokinetical behaviour to a large extent. The equilibrium between the free and bound part of the radiopharmaceutical enables the transfer to tissues and organs at any time, where it can be bound, metabolized or eliminated.
Until now some aspects of transport and binding of radiotracers in blood have been considered.

Table 12.7. Relationship between serum protein binding and some parameters of organ distribution

Compound	Serum protein binding	Maximal liver/bowels uptake	Urinary excretion 30 min p. i.
Tc-99m-diethyl-IDA	35 %	80–85 %	10–14 %
Tc-99m-4-methyl-IDA	10 %	50–54 %	30–35 %
Tc-99m-p-brom-IDA	60 %	80–85 %	8–10 %
Tc-99m-2,4,6-trimethyl-IDA	14 %	80–85 %	12–14 %

After intravenous injection the efflux begins immediately into the other distribution spaces. Differences in structure of capillaries in single organs lead to differences in the distribution of radioactive drugs.
Generally transport in the organism includes passage through cell membranes. Important properties of penetration are, as mentioned above: solubility in water and lipids, the particle size and molecular weight respectively, and adsorption to carrier substances.
As a result of these properties, different chemical and physiological affinities exist between the radiopharmaceutical and the tissue. These affinities have a remarkable influence on tissue uptake and retention. The *membranes* are of major *importance* for any kind of pharmacokinetical behaviour. These membranes are the place of different transport systems in the organism. Figure 12.3. shows a schematic representation of the membrane structure. The composition of membrane constituents strongly varies from organ to organ and tissue to tissue. This fact excludes a homogeneous distribution a priori.

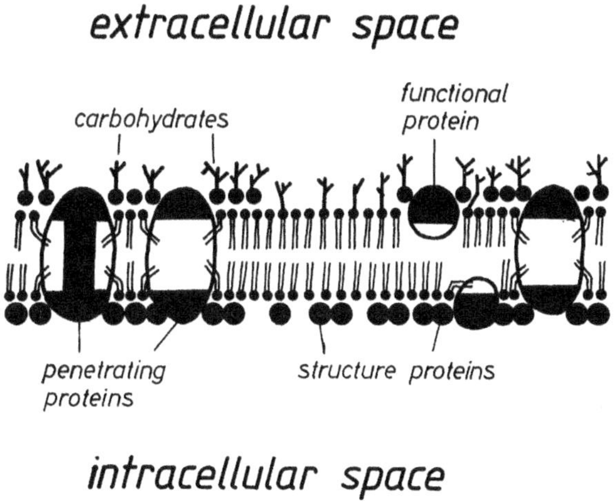

Fig. 12.3. Model of membrane structure

A sufficiently specific uptake is important for the clinical use of radioactively labelled substances. The quick elimination of remaining radioactivity from the body is preferred in order to give a high target-to-non-target ratio.
In the past we performed a large number of biological investigations with Tc-99m-diethyl-IDA, which enables the assessment of the excretory hepatocyte function. As we carried out labelling with a home-made preparation, it was neccessary to perform biological quality control in comparison to the commercial kit before clinical use. Moreover, we tested further derivatives of this class of compounds. The results of investigations into organ distribution with these substances may serve as an example for the characterization and the improvement of radioactive drugs, respectively.
In general the first step is the optimization of the labelling conditions and the choice of method to check in vitro stability. For analytical purpose of Tc-99m-IDA-derivatives we used paper electrophoresis. The pH-value, ligand concentration, substance of chemical reduction and reaction time are some parameters, which are important for the preparation. Each preparation is electrophoretically tested parallel with the biological test. Besides the Tc-99m-diethyl-IDA-complex and a little quantity of pertechnetate, two other complexes were obtained at a low percentage. The second step is biological quality control in laboratory animals. The animals were killed by decapitation, the organs dissected, weighed and the radioactivity measured. The calculation of radioactivity per g tissue and the whole organ followed. Table 12.8. demonstrates the possible variations in organ distribution dependent on the time-interval between labelling and application of radiopharmaceuticals. In other words, the storage time may influence the chemical composition and the biological behaviour of labelled substances.
The urinary excretion rate is one important quality criterion for hepatobiliarily eliminated drugs. Concentration in blood and urine is lower 2 and 4 hours after preparation. From these observations the necessity was concluded to take into account the constant time intervals between preparation and injection of radioactive drugs in those experiments to avoid errors. For any experiment 50 to 60 animals on average are needed if 5 animals for each time interval are used. In the case of Tc-99m-IDA-derivates we concluded that they could be used between 60 and 120 minutes after preparation. These examinations have to be performed at the beginning of pharmacokinetical studies, especially with Tc-99m-labelled complexes.

Table 12.8. Organ distribution of Tc-99m-diethyl-IDA 30 min p. i. in dependence on storage time (cpm/organ in % of dose)

Time p. p. (min)	Liver/bowels	Kidney	Blood	Urinary excretion
30	66	1.1	1.6	16
60	75	0.9	1.1	12
120	80	0.7	0.9	10
240	81	0.5	0.6	10

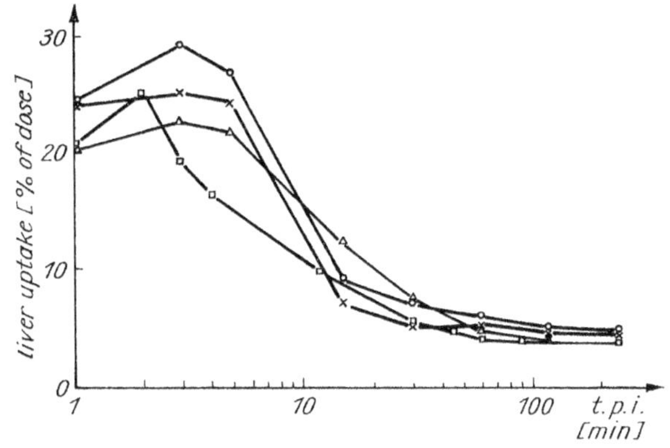

Fig. 12.4. Liver uptake of Tc-IDA-derivates in ICR-mice (cpm/g in % of dose)
○——○ Tc-99m-trimethyl-IDA
×——× Tc-99m-p-brom-IDA
△——△ Tc-99m-4-methyl-IDA
□——□ Tc-99m-diethyl-IDA

The maximal liver uptake of Tc-99m-diethyl-IDA was determined about 2 minutes p.i. (Figure 12.4.). The same radioactive concentration was measured in the liver and kidney 1 and 2 minutes p.i. (cpm/g), an indication, that not only a transport mechanism, specific to the liver, is responsible for elimination from blood. It can also be possible that one fraction, present in the preparation, is exclusively eliminated through the kidney. In comparison to Tc-99m-diethyl-IDA the liver-kidney-quotient of Tc-99m-p-brom-IDA and Tc-99m-trimethyl-IDA is higher by the factor 2 and lower by the factor 2 for Tc-99m-4-methyl-IDA. The urinary excretion amounts to 8—12 % of the dose for the first compounds, for the last one 30—40 %. These examples show that a discrete alteration of the molecular structure can make possible a more specific organ uptake.
The question of metabolization of any radioactive drug can be analyzed on the one hand by analytical-chemical methods after organ passage, and on the other by observing the biological behaviour after reinjection of the radioactive substances eliminated with the body fluids.
For example electrophoresis after body passage of Tc-99m-diethyl-IDA shows the Tc-99m-diethyl-IDA-complex by examining the bile; examining the urine, besides this complex another one with lower electrophoretical mobility was found.
The organ distribution of the complex contained in bile fluid after reinjection is about the same as before (Table 12.9., 12.10.). This recirculated Tc-99m-IDA-complex shows only a little change in its biological behaviour. On the contrary, the reinjection of urine yielded a different distribution pattern. After reinjection of urine and bile Tc-99m-trimethyl-IDA shows an identical organ distribution, while the experiments carried out on Tc-99m-brom-IDA reveal a metabolization of the complex.

Table 12.9. Maximale liver/bowel uptake of Tc-99m-IDA-derivates (cpm/organ in % of dose) in ICR-mice, 30 min p. i.

Tc-99m-R-IDA	First passage	Reinjection urine	Reinjection bile
R: 2.4-diethyl-	70	40	78
4-methyl-	55	43	47
p-brom-	76	50	62
2.4.6-trimethyl-	79	76	78

Table 12.10. Maximale urinary excretion rate of Tc-99m-IDA-derivates (% of dose) in ICR-mice, 30 min p. i.

Tc-99m-R-IDA	First passage	Reinjection urine	Reinjection bile
R: 2.4-diethyl-	14	30	10
4-methyl-	30	36	38
p-brom-	8	28	18
2.4.6-trimethyl-	9	13	12

Before introducing a new compound into clinical application, the biological investigations mentioned above have to be performed in another species. Then clinical testing can start with a limited number of volunteers.
The experimental conditions of the trial programme have to be accomodated to the special clinical problem. Some further methods which are applied to the study of radiopharmaceuticals at departments of nuclear medicine can only be mentioned. The measurement of whole-body retention serves above all the calculation of the biological half-life and is the basis for calculating the radiation dose. The course of time of biological processes at the cellular level can be observed with the autoradiographic technique. It makes possible an exakt localization of radioactive substances in organs, tissues, cells and cell structures. The whole-body-animal autoradiographic method provides a clear description of the relative concentrations in organs, which can be seen in the histological slice. The same but not so detailed statement renders possible the performance of functional scintigraphy.
For testing and developing radiopharmaceuticals it is very important to gain more information on physiology and to explore variations and impairments of single organ functions under pathological conditions. Therefore, studies into easily injured animals would be an essential step to extend the biological testing of radioactive drugs.
Also sensitive reference methods are necessary to provide information on small pathological variations. It should be considered that the applied nuclear medical methods, which are based on the application of radioactive drugs, should not be less accurate than conventional laboratory techniques.

References

1. *Struck, H.:* Experimentelle Medizin. Thieme, Stuttgart 1973.
2. *Glenn, H. J.* (ed.): Biologic Applications of Radiotracers. CRC Press, Inc. Boca Raton, Florida.
3. Technical Reports Series No. 193: Laboratory Training Manual on the Use of Nuclear Techniques in Animal Research. IAEA, Vienna 1979.
4. *Eckelmann, W. C.:* Receptor-Binding Radiotracers. CRC Press, Inc. Boca Raton 1982.
5. *Cox, P. H.* (ed.: Progress in Radiopharmacology. Vol. 1 and 2. Amsterdam–New York–Oxford 1979/1982.
6. *Colombetti, L. G.* (ed.): Principles of Radiopharmacology. Vol. I–III. CRC Press, Inc. 1979.
7. *Spencer, R. P.:* Radiopharmaceuticals–Structure–Activity–Relationships. Grune and Stratton, New York–London–Toronto–Sydney–San Francisco.
8. *Subramanian, G., B. A. Rhodes, J. F. Cooper, V. J. Sodd:* Radiopharmaceuticals Society of Nuclear Medicine, Inc. New York 1975.

Chapter 13

Principles, Problems and Trends in Radiopharmacology

Bernd Johannsen

Employing Hevesy's concept of radioactive indicators nuclear medicine plays an important role in an increasing number of clinical situations. The principle of the dynamic state of body constituents which is regarded as the fundamental concept, could be derived from classical radiotracer studies already during the 30's and 40's. After an intermediate period of concentration on primarily static morphological information, there is at present a reorientation towards the tracer principle with emphasis on regional physiology and biochemistry.

The goal, to visualize the structure and function of the human body noninvasively in the form of spatial and temporal distribution patterns of the indicators, has induced a steadily growing demand for new *radiotracers*.

The purpose of this Chapter is to provide sufficient background information on principles and problems of radiotracer design to make possible evaluations of advances and trends in the rapidly increasing radiopharmacological literature.

1. Radiopharmaceuticals and drugs – different subjects?

From the legal point of view, a radiopharmaceutical (termed "radioactive drug" in the national Drug Act of the GDR) is regarded as a drug just like any other kind of drug. Therefore, differences which have to be stressed here do not concern this aspect but are based upon the following considerations:

1. Unlike drugs administered, with few exceptions, radiopharmaceuticals are intended to trace a process without altering it in order to produce therapeutic effects at appropriate concentrations. Consequently, the amount administered is very small, e.g. at the picomole level. The value of the tracers does not depend on their concentration. It could sometimes even be improved by reducing the concentration to a minimum (using carrier-free radionuclides), for example in receptor-mapping or special cases of tumor localization. The few radiopharmaceuticals used for therapeutic purposes are also without any desired effect caused by the amount of substance.
2. While current pharmaceutical research can make use of the enormous variety of organic syntheses, which allows that almost any substance desired can be prepared, the possibilities in radiotracer syntheses are limited because of the restriction to few radionuclides suitable for rou-

tine use in nuclear medicine Fig. 13.1. and 13.2. One of the major problems in the synthesis of radiotracers is the absence of convenient gamma-emitting radionuclides of the key bioelements carbon, hydrogen, oxygen and nitrogen. Increasing awareness of biochemical aspects of radiopharmaceutical design has stimulated the syntheses of organic molecules with carbon-11 (half-life of 20 minutes) despite the economic, technical and further problems involved.

A striking feature of present-day radiopharmaceuticals is the predominance of inorganic substances, in particular such compounds biochemists and pharmacologists are not familiar with. This is most evident in compounds of Tc-99m, the most widely used radionuclide in nuclear medicine.

Hence, radiopharmaceuticals differ from therapeutically used drugs both in the medical purpose desired and in the chemical nature or synthesis strategy.

2. The objective of radiopharmacology

The value of a radiotracer depends on its biodistribution under physiological and pathological conditions. To understand or even to predict this very biological behaviour—particularly of "substrate--specific" radiotracers according to the classification by Eckelman and Reba /1/—profound knowledge is necessary about the relationship between the chemical structure or the properties of the radiotracer and the process involved in biodistribution such as spatial distribution, plasma protein-binding, transport, binding inside the cell, metabolism and excretion.

As stated by Colombetti in 1979 /2/, radiopharmacology is the science which studies the chemical properties of radiotracers (radiopharmaceuticals) and

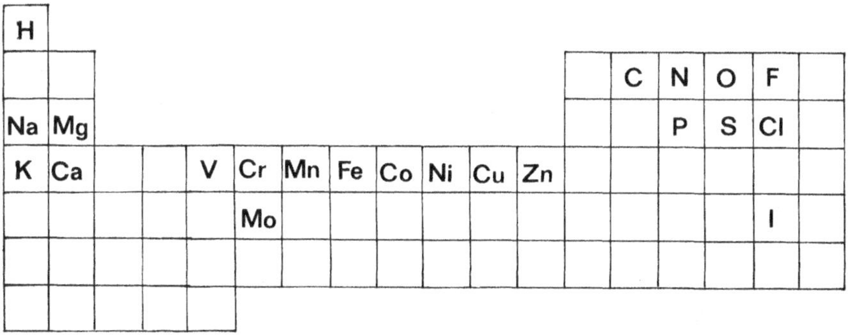

Fig. 13.1. Bioelements in the periodic table of elements

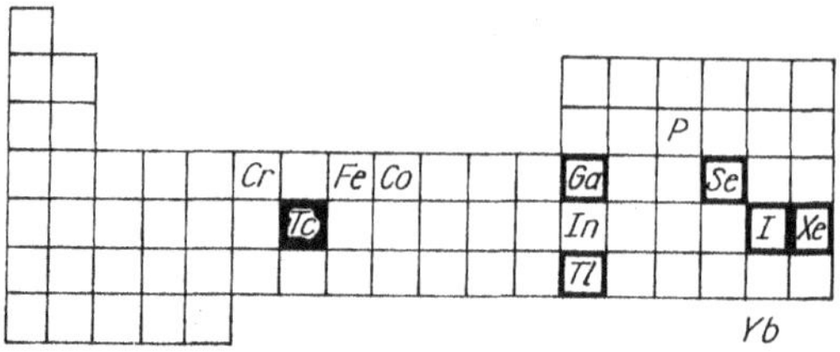

Fig. 13.2. Elements mainly used as radionuclides in diagnostic nuclear medicine

their interactions with living organisms. Defined in this way, radiopharmacology is an interdisciplinary branch that integrates appropriate components of chemistry (in particular radiochemistry, pharmaceutical chemistry, biochemistry with emphasis on bioinorganic chemistry), pharmacology and physiology.

Of course, the well-established pharmacology and the emerging discipline of radiopharmacology have a number of features in common. The parallel should, however, not be carried too far because of the important differences between a "true" drug and a radiopharmaceutical as discussed above. We can scarcely find radiopharmacological papers dealing with the pharmacology of radiotracers in any journal of pharmacology but in periodicals of nuclear medicine and radiopharmacy as well as in proceedings of symposia on radiopharmacology /3, 5/.

In the early stages of radiopharmaceutical research efforts were mainly focussed on discovering new classes of radiopharmaceuticals rather than on selecting the most promising candidates within a class of compounds or on trying to clarify the biological behaviour on a causal basis. Some radiotracers could be introduced for obvious reasons, e.g. I-131 iodide, labelled dyes and labelled colloids of particles, respectively. In most cases, the radioactive substance applied has no counterpart in normal use.

This is obvious in the case of artifical substances such as Tc-99m-complexes. Substitution in critical positions within the molecular framework of physiological substances or drugs by a "foreign" element also makes it difficult to explain the mechanisms of biodistribution. At present the situation is changing. The growing demand for specific radiotracers and the increasing awareness of the progress in molecular pharmacology and biochemistry have produced profound studies into the two fundamentals of *structure-biodistribution relationship,* into

– chemical properties of radiotracers,
– mechanisms of biodistribution.

Elucidation of the relationship itself is still in the early stages. The potential value of the use of quantitative structure-biodistribution relationship (QSAR) for selecting the most promising candidates within a class of homologous compounds has of course been recognized and demonstrated recently. The main problem in radiotracer design consists in finding new *major structures* that can serve as a starting point for structural variation or optimization. As in the field of drug design, the application of biochemical or any other rationales to the discovery of new leads is of particular interest. Even if we were able to create a lot of potential structures relevant to diagnostic use by means of biochemical, pharmacological or any other working hypotheses, most of these structures could not be employed directly because of the limitations in radiopharmaceutical synthesis. Hence, we are faced with the additional problem of deriving adequate or even "artificial" structures from the real or putative "natural" lead structures. The term "analogues concept" has to be introduced in order to paraphrase a guideline suggested in this situation.

3. The "analogues concept": Imitation of a non-radioactive lead by the potential radiotracer

The concept becomes already evident in the simple case of inorganic ions as illustrated in the following example Fig. 13.3. Owing to the role the $Na/^{+}$ K^{+}-ATPase pump plays in the viable myocardium, radioactive potassium ions were suggested for myocardial imaging. Because of the poor characteristics of the radioisotopes of potassium, however, their use is not favourable. If nevertheless the potassium pump should be utilized, other ions which might mimic potassium are desired. The permissible variation relates to the charge and size of the ion or its hydration sphere. The application of the "analogues concept" along this line led to Tl-201 thallous chloride, which was introduced in 1975 by Lebowitz and represents the agent of choice among potassium analogues.

Thus, the "analogues concept" enables us to "translate" target molecule structures of "foreign" substances such as the toxic thallous ion in the example discussed above. At present background knowledge is often not sufficient, even in this simple case of potassium transport into the cell. A true interrelationship between the uptake of K^{+} and Tl^{+} has not been proved until now but called in question. This uncertainty increases with the complexity of the molecules of interest.

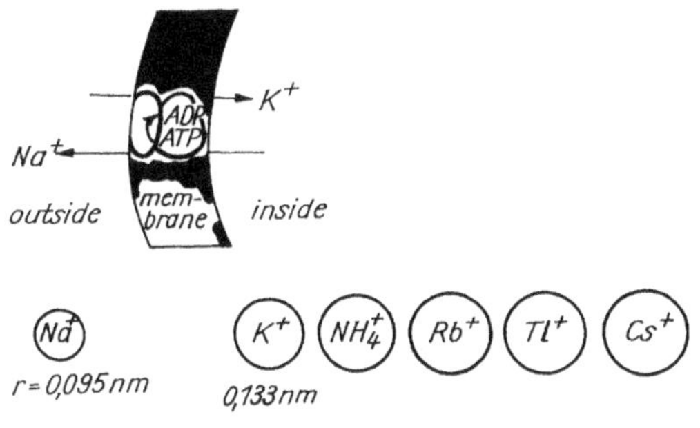

Fig. 13.3. Potassium ion analogues as substrates of the potassium pump

Meanwhile efforts have been focussed on replacing Tl-201 by a Tc-99m agent. The chemistry of technetium precludes the use of a simple cation which resembles the potassium ion.
Fortunately, various organic cations, such as some cardio-selective drugs, substrates of enzymes, neurotransmitters, etc. accumulate in the myocardium. Therefore, the search for cationic Tc-99m myocardium agents may prove rewarding. Several authors succeeded in preparing cationic technetium complexes. The cationic character itself proved to be insufficient. Deutsch et al. prepared a special type of cationic Tc-99m complexes with a + 1 formal charge. According to animal studies of the complexes investigated so far, $Tc(DMPE)_2Cl_2^+$ is promising. Despite the sensitivity of biodistribution to subtle change in molecular configuration and uncertainty concerning the precise mechanism, it is most likely that putative lead structures (physiologically important organic compounds) can be simulated by molecules which belong to quite different types of chemicals, e.g. coordination compounds.
The potentialities of radiotracer design depend on the variety involved in the chemistry of those elements which suitable radionuclides belong to. For this reason, some chemical aspects of radiotracers will be discussed with special emphasis on technetium chemistry. Radioactive gases, colloids, particles and cells will not be considered here. The same applies to new "nonspecific" tracers of ultra-short-lived generator nuclides for first pass studies.

4. Inorganic tracers

The group of inorganic radiotracers comprises simple ions and complexes. Most of them do not occur in the human body. Like the ions of alkali metals, the alkaline earth ions are primarily characterized by their ionic size and hydration energy. Differences in these parameters are very important in biological systems because the biological system differentiates at definite positions in the periodic table of the elements strictly, e.g. between Na^+ and K^+ or Mg^{2+}. Thus, similarities in ion parameters determine the choice of substances, if only analogues are taken into account for radiotracers or, on the other hand, lead to an understanding of the in vivo behaviour of ionic radiopharmaceuticals.
The replacement of Ca^{2+} radioisotopes by *Sr-85* or *Sr-89* ions for special indications is a case in point just like the interpretation of the osteotropic properties of radiolanthanides or Ga^{3+} as an ionic replacement at Ca^{2+}-binding sites. Much of the application of radioisotopes of *gallium* and *indium* is connected with the similarity of these Group III elements to the transition element iron. Their M(III) ions mimic high-spin iron(III) in complex formation. Therefore, the high affinity to transferrin can be clearly understood from the thermodynamic point of view, the formation of transferrin complexes is a disadvantage, if less stable complexes are to be used not only for blood-pool imaging. Considering the stability constants, Ga^{3+} and In^{3+}

complexes with O-donor ligands (e.g. citrate) or O,N-donor ligands (e.g. EDTA, DTPA) are expected to be converted into the transferrin complex after injection, thus exhibiting the same or similar biodistribution. Kinetic phenomena, however, can control the ligand exchange or biodistribution. This is useful for the EDTA complexes which proved to be kinetically inert. Possibilities of overcoming the transferrin barrier consist on the one hand in developing EDTA-containing bifunctional agents or, on the other hand, in exploiting such ligands which are able to form extremely stable complexes. Such ligands have been found (desferrioxamine and tricatecholamide moieties).

Further research is progressing in this area. The most fascinating complexes relevant to radiopharmaceuticals are those of *technetium* because of their very complicated and profitable chemistry. We are faced with the fact that Tc does not simulate any bioelement. Intensive research has been focussed on the elucidation of the chemistry of technetium both in model studies with the long-lived Tc-99 isotope and under "no-carrier-added" conditions of Tc-99m (10^{-7} mol · l^{-1} or less) and on the biological behaviour of defined complexes /e.g. 5/.

In contrast to only "Tc-tagged" radiopharmaceuticals in which Tc is attached to large molecules without significantly interfering with properties of the native compound, "Tc-essential" complexes are "tailored" compounds with special molecular configurations and molecule parameters.

Technetium forms complexes at every oxidation state from − 1 to + 7. Apart from pertechnetate itself, Tc(V), Tc(IV), Tc(III) and Tc(I) are of particular interest here. These central atoms are stabilized by ligands forming complexes of defined geometry, charge, size, solubility, reactivity, etc.

As can be seen in Figure 13.4., the important key to interpreting the behaviour of the complexes is the structure of the entire complexes rather than that of the free ligands. As a result of complex formation, the reactivity of the original ligand is altered insofar as functional groups are completely or partly involved in binding the metal. There is an unavoidable increase in size and more or less alteration of the charge. Different types of complexes can be formed depending on the oxidation state of Tc, type of ligands, the

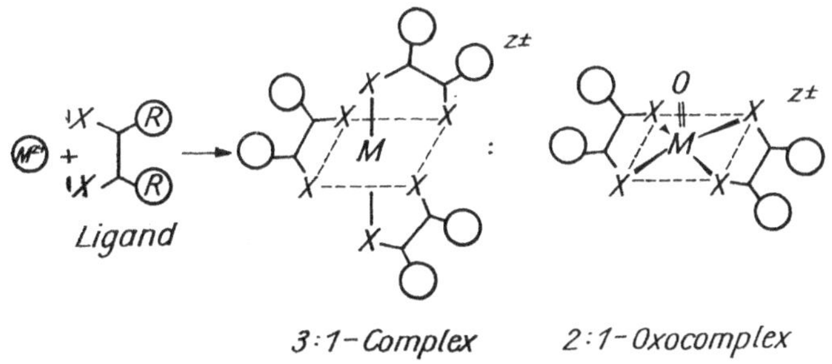

Fig. 13.4. Schematic representation of complex formation

coordination number, etc. Suitable ligands involve compounds possessing functional groups such as hydroxyl, carboxyl, amino, mercapto, phosphonate in an appropriate molecular arrangement. The most commonly used ligands are multidentate ligands.
Without going into detail, it may be stated that the oxidation state of technetium seems to influence in some systems the biokinetics due to different cores (e.g. $Tc^{V}O^{3+}$, $Tc^{IV}OOH^{+}$, Tc^{3+}), geometries and reactivities. While most Tc-complexes used as radiopharmaceuticals represent hydrophilic anions, there is a growing interest in cationic and neutral complexes. The design of Tc-radiotracers should also take into account the ratio of Tc to ligand.
There might be a need for complexes with an overall asymmetrical shape thus perhaps resembling that of physiological substrates or organic drugs. Considering the role cationic substrates and drugs play in biological systems, cationic technetium complex ions may enrich the radiotracer assortment as far as they are able to mimic cationic lead structures.
The preparation of neutral, lipophilic Tc-tracers capable of passive diffusion across lipoidal membranes has been the purpose of recent studies.

5. Organic tracers

In principle radiotracer design is no longer limited to inorganic or few organic substances. The increasing availability of the short-lived carbon-11 (half-life of 20 min) leads to a synthesis variety which seems to be unlimited provided there are fast procedures. The restriction valid in non-isotopic labelling does not exist. We are no longer dependent on the "analogues concept". The specific activity can be kept sufficiently high to satisfy the requirements for studies at a very low tracer level. A large number of organic compounds have been prepared in carbon-11 labelled form. Due to economic and other problems there is a deep gap between great achievements attained in research and clinical routine application.
A very interesting class of compounds are non-isotopic-labelled organic molecules. These are compounds labelled with an isotope of foreign elements not normally present in the unlabelled substance.
Since this kind of labelling represents a definite substitution within the parent molecule Fig. 13.5., its effect can be assessed with the aid of rules of known eletronic, hydrophobic and steric substitution effects. The term *bioisosterism* seems to provide much room for substrate-specific radiotracers.

Fig. 13.5. Isotopic and non-isotopic labelling of organic compounds represent distinct substitution (X = substituent) at the carbon skeleton

Halogenation is predominantly performed with radioisotopes of iodine. It involves relatively large substitution effects and the problem of dehalogenation in vitro and in vivo. Substitution with bromine-75 offers an alternative, if a stronger C-X-bonds is required. The labelling with fluorine-18 is of particular interest. F-18 is favourable because of its longer half-life of 110 minutes compared with C-11. Fluorine is of comparable size to hydrogen but the chemical reactivities are very different. Depending on the substitution position this fact is more or less important. Enzyme substrates may be converted into inhibitors.

Besides C-F in place of C-H, a fluorine atom can also be introduced as the analogue of a OH-group. Since the fluorine atom can only act as hydrogen-bond acceptor instead of a donor and acceptor, transport based on H-bonding may be disturbed. Other foreign labelling has been also suggested. Bioisosteric or drastic substitutions have been performed with radioisotopes of selenium (bioester to sulphur), tellurium, arsine (replacement of phosphorus), etc.

6. Mechanisms of biodistribution

If there is any relationship between the structure of the radiotracer and its biodistribution, we have to consider which mechanisms of distribution, localization and excretion fit with the structures or molecular parameters discussed above.

Apart form the initial flow-dependent distribution, the processes can be very complex—in particular as far as the substrate-specific radiotracer are concerned—and they have been little understood until now.

To outline some factors, different suggestions concerning the mechanism of the uptake of Ga-67 (administered as gallium citrate) should be mentioned: hyperpermeability of the plasma membrane of tumor cells, isomorphous replacement at cationic binding-sites in tumor tissue, bindings with transferrin, lactoferrin or other receptors and binding to siderophores. The explanation of the mechanism involved in the uptake of Ga-67 citrate into malignant and normal soft tissues is one of the most controversial subjects of radiopharmacology. Multiple factors are probably involved.

Figure 13.6. shows plasma protein binding, two basic pathways which might be involved in the penetration into soft-tissue tumors, normal soft tissues and inflammatory lesions and intracellular binding.

Regardless of whether or not gallium actually enters the cell in the manner suggested, the model emphasizes the following processes which are also relevant to other radiotracers: plasma protein binding, transmembrane transport and intracellular reactions.

Plasma protein binding: The theoretical concepts of drug binding by serum proteins as well as the effect of binding on the clearance of drugs are now well defined and apply to radiotracers too, if their very low concentrations are taken into account.

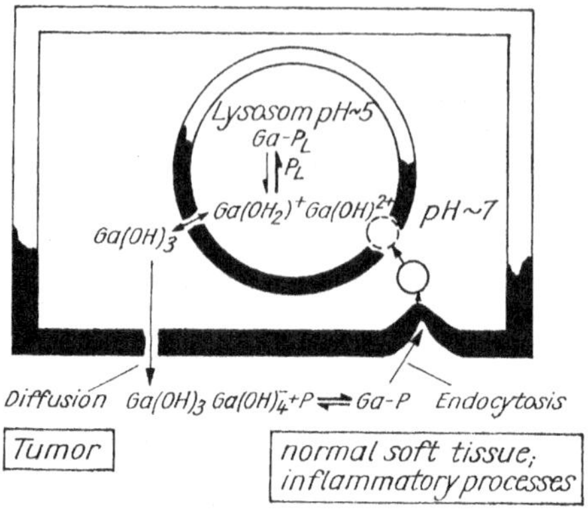

Fig. 13.6. Schematic model of the uptake of Ga-67 citrate (Hayes et al. 1982)

7. Transmembrane transport

Generaly, permeating substances may cross the membrane in different ways as illustrated in Figure 13.7. Non-carrier transport either via pores or due to high lipophilicity of the molecule always proceeds in the direction of the concentration gradients. Many substances can be transported by carrier in more or less specific transport systems. Active transport processes make possible up-hill transports, such as K^+ pumping into muscle cells (see Figure 13.3.) or iodide uptake into the thyroid gland. Penetration of various substances across membrane depends both on the properties of the solutes and on the characteristics of the membrane.

Passive transport through lipoidal membranes such as the intact blood-brain barrier is possible for substances with favourable lipid/water partition co-

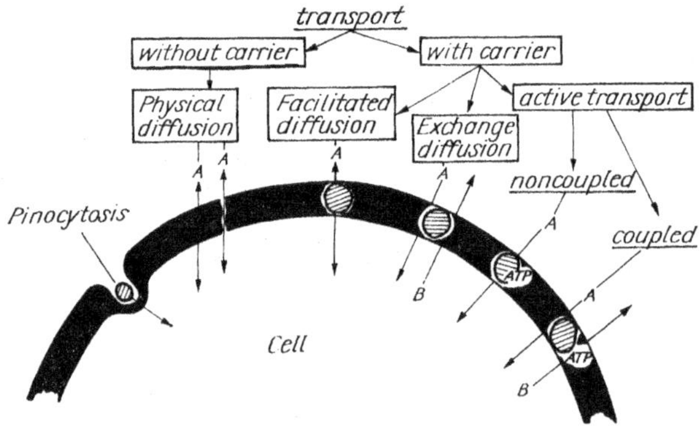

Fig. 13.7. Transmembrane transport – an important factor in radiotracer design

efficients. Weak bases with suitable pK values can be used as pH-dependent radiotracers (*pH-shift agents*) which are neutral and lipid-soluble at blood pH (pH 7.4) and diffuse freely into the cells.

If the intracellular pH is significantly lower than the blood pH, the bases become ionized and thus trapped within the cell because the charged molecules cannot readily diffuse out of the cell. Many tissues have a low intracellular pH including actively metabolizing tissues such as brain, heart and some tumors.

According to the model of Ga-67 uptake into tumor cells shown in Figure 13.6., the penetration mainly occurs by means of passive diffusion of free uncharged $Ga(OH)_3$ across the plasma and lysosomal membranes, followed by binding of the ionized species produced at a lower pH.

The pH-shift agents designed for perfusion studies of the brain include Se-containing diamines (PIPSE, MOSE) and radioiodinated amines. Optimization of the latter class led to N-isopropyl-p-(I-123)iodoamphetamine with a first-pass extraction efficiency of almost 100 %. Phagocytosis and pinocytosis make it possible to transfer macromolecules and particles, respectively. Localization of radiotracers in the RES or macrophage system has been utilized for a long time. It is possible that selective pinocytosis plays a role in the localization of radiotracers in certain tumors.

According to the model in Figure 13.6., phagocytosis might be responsible for the uptake of Ga-67 into normal and inflammatory tissues.

Carrier-mediated processes, especially active transport, enable the cell to exchange substrates to concentrate nutrients and to remove waste products. These vital functions are maintained by rather specific carriers. Radiotracers to be designed on this basis must be able to compete with the natural permeants. Therefore, the structural requirements are dictated by the tolerance of the carrier system. Radiotracers assumed to be transported in this way are for example TcO_4^- because of its similarity to iodide with respect to the iodide pump, I-131-orthoiodohippurate in the renal transport system for organic anions (hippurate transport system) or radioiodinated dyes and some Tc-99m complexes in the anion transport system of the liver. The design of Tc-99m complexes capable of simulating organic anions handled in both excretion mechanisms is a great challenge. The more difficult task is to find a *substitute for I-131-iodohippurate* compared to synthesizing hepatobiliary Tc-99m agents.

Although structural similarities are not immediately apparent between the two structures compared in Figure 13.8., they exist regarding the charge, amide group, etc.

Starting from Tc-DADS, rapidly excreted through the kidneys, several derivatives of this complex have been synthesized. The addition of a carboxylate group to the ethylene bridge portion results in two isomeric components. One of them is nearly equivalent to I-131-orthoiodohippurate as regards rate and specificity for renal excretion.

Due to the tolerance of the hippurate transport system in the kidney, even a Tc-complex of "functionalized" PAH cannot be obviously ruled out as recent studies suggest.

The design of *hepatobiliary Tc-99m complexes* seems to be easier and is more

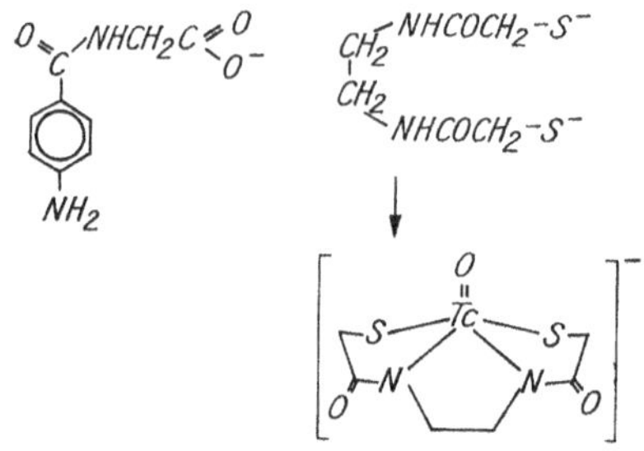

Fig. 13.8. Comparison of the structures of ortho-iodohippurate and Tc-DADS (Tc-N,N'-bis(mercaptoacetamido-ethylenediamine)- a potential radiotracer lead structure for a renal tubular function agent

advanced than that of renal function agents. There are known rules concerning structural requirements for biliary excretion which can be applied to technetium complex synthesis. Meanwhile various types of Tc-complexes have been extensively investigated including QSAR analysis. One of the classes represents Tc-complexes of Schiff-bases of pyridoxal and amino acids. The putative structure of the Tc-pyridoxylidenephenylalanine complex (Figure 13.9.) shows the required high molecular weight, polar groups and ring systems in different planes.

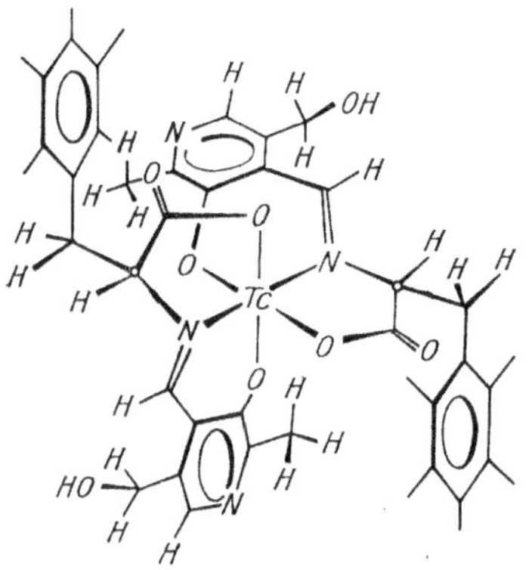

Fig. 13.9. Possible structure of the Tc-pyridoxylidenephenylalanine complex as an example of hepatobiliary agents (Kato-Azuma 1980)

Another class of Tc-99m hepatobiliary agents originated in Tc-99m-HIDA (Figure 13.10., 13.11.). Optimization of this class led to agents with some different specificity and/or sensitivity in patients with normal and reduced hepatocyte function. Reduction of the imine link in the Schiff-bases complexes to stable amine linkage, recently performed by Kato-Azuma, opens a new class of agents which are promising because of rapid blood clearance, fast hepatobiliary transit, low urinary excretion and strong resistance to serum bilirubin.

Tc-99m-PHMT (PHMT = N-pyridoxyl-DL-5-methyltryptophan) is hitherto the most suitable candidate.

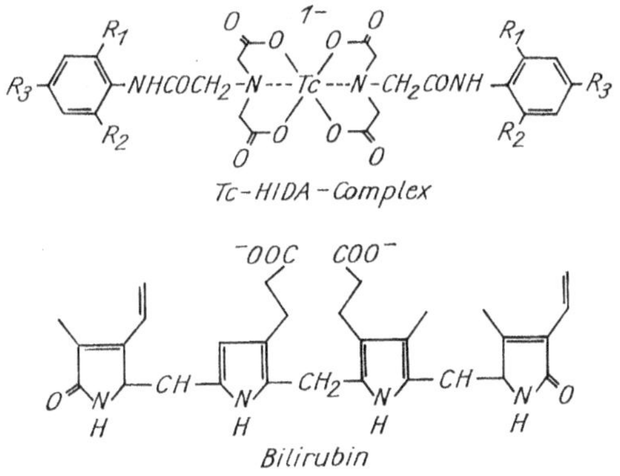

Fig. 13.10. Comparison of the structures of the physiological bilirubin and artificial Tc-HIDA-complexes

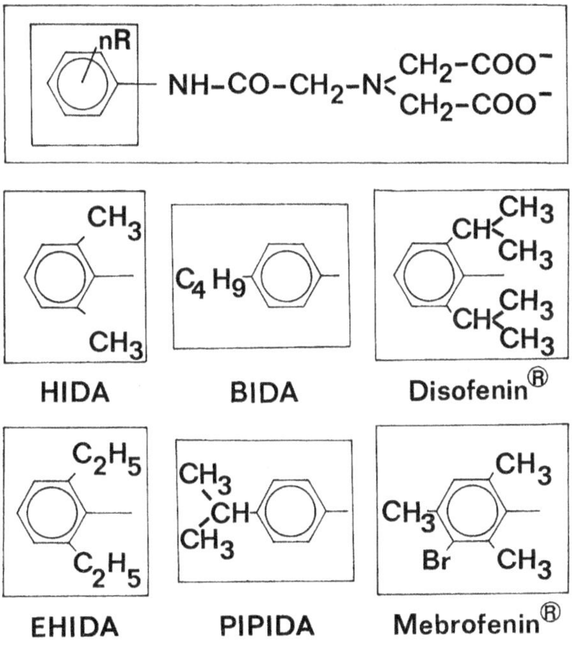

Fig. 13.11. Some important Tc-HIDA complex homologues obtained by variation of the alkyl substituents of the phenyl ring

Promising systematic design of myocardial imaging tracers such as Tc-DIARS, Tc-DMPE and $Tc(CNR)_6^+$ (*Deutsch* et al. 1981, *Jones* et al. 1982, *Robbins* et al. 1984) has been started on the basis of still unknown cation uptake mechanisms. The main types of cationic Tc-species which represent starting points in this context are schematically compiled in Figure 13.12.

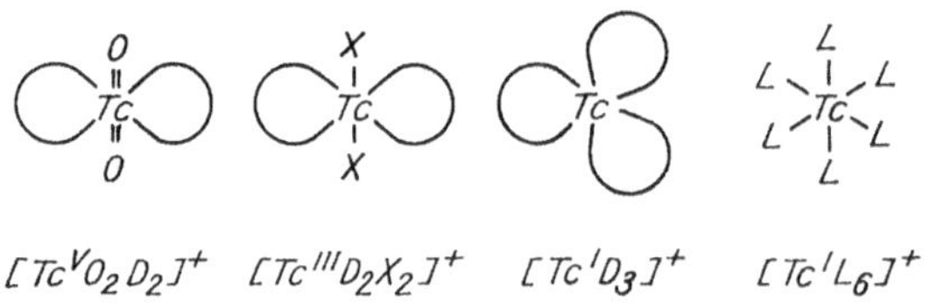

Fig. 13.12. Classes of cationic Tc-complexes used in radiotracer design (the 3 types on the right are relevant to myocardial imaging)

8. Selected biochemical pathways

If biodistribution based on a combination of transmembrane transport and intracellular reaction (metabolic pathway or binding at specific sites) is desired, the structural requirements for radiotracers become stricter. There is hardly a chance of Tc-complexes to pass the enzymatic sequence of a pathway. Moreover, to my knowledge, it has not yet been studied whether any Tc-complex or coordinated ligands can be attacked by enzymes. Hence, metabolic studies have been confined to the domain of isotopically labelled substrates or their precursors (C-11 labelled compounds) or certain non-isotopically labelled substrates which follow a selected biochemical pathway or part of it.

The problem is to identify a reaction or reaction sequence which occurs to a greater extent or at a faster rate in the desired target tissue as compared to the surroundings. Under this premise radiotracer design has been focussed on the metabolism of amino acids, in particular tyrosine (catecholamine, melanine precursors), nucleosides, biogenic amines and especially glucose and fatty acids.

Under aerobic conditions, free fatty acids are primary substrates for normally perfused myocardium. As a result of high utilization by the heart, rapid uptake from the blood and minimal concentration in the normal lung, labelled fatty acids offer an excellent potential for the study of regional myocardial metabolism.

Apart from non-modified substrate (C-11 palmitate), various approaches to non-isotopically labelled substance have been suggested, including terminally radioiodinated long-chain acids and labelled phenyl-substituted acids or fatty acids with a heteroatom within the alkyl chain in order to avoid complete catabolism, thus reducing washout from myocardium.

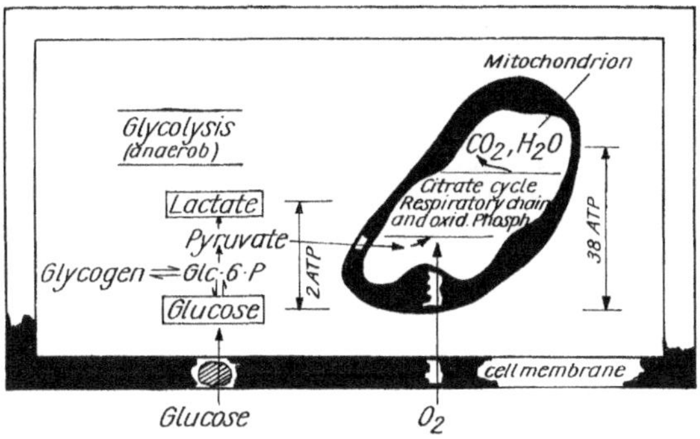

Fig. 13.13. Scheme of regional glucose metabolism in the cell

Fore more detailed consideration see e.g. the article by Feinendegen et al. /7/. Regional brain glucose metabolism is another lucrative target. D-glucose is the energy source of the brain cell (Fig. 13.13). After facilitated diffusion into the cell, decomposition of glucose starts with a conversion into glucose-6-P catalyzed by hexokinase. This initial transport and first step as well as the entire metabolism can be traced by means of C-11-labelled glucose. Regarding the complexity of glucose metabolism, it appears to be favourable to isolate transmembrane transport either together with the following step or not.

Fortunately, this is possible through modified labelled substrates (Fig. 13.14). 2-deoxy-D-glucose can be considered a lead structure. The structural alte-

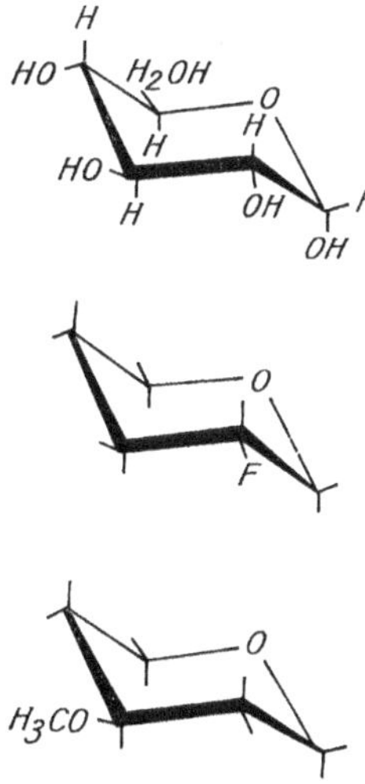

Fig. 13.14. Systematically modified structures for studying glucose metabolism:
C-11-labelled glucose – the unaltered substrate
F-18-2-deoxy-2-fluoro-D-glucose for metabolic trapping
3-C-11-methyl-D-glucose – tracer of glucose transport

ration within this molecule is tolerated by hexokinase. The membrane permeability of the enzymic product of deoxyglucose phosphate is low and, furthermore, cannot enter into the subsequent metabolic steps. Hence, at high hexokinase and low phosphatase activities the product will be metabolically trapped. Substitution of C-F in place of C-H at C-2 in deoxyglucose (i.e. C-F in place of C-OH in glucose) has no negative effect. Therefore, F-18-2-deoxy-2-fluoro-D-glucose allows *metabolic trapping*. The labelled compound favourably differs from glucose in the renal reabsorption mechanism. The suitability of this tracer is supported by several factors /8/. Another approach in this field is the development of a radiotracer for glucose transport only.

Stöcklin et al. have suggested 3-C-11-methyl-D-glucose for this purpose. Figure 13.14. shows the differences between the structures of radiotracers.

9. Intracellular binding

Because of the outstanding role of *osteotropic radiotracers* in nuclear medicine, skeletal uptake will be briefly discussed. Despite numerous investigations into factors involved in skeletal uptake, the mechanism itself remains poorly understood. The Tc-99m complexe are simultaneously taken up by bone and rapidly cleared through the kidneys. The movement from the blood to bone is assumed to be passive diffusion through the clefts in the capillary wall into the ECF.

There is no consensus of opinions whether renal clearance or bone uptake is the primary factor affecting scintigraphic quality. The tracer is likely to be adsorbed onto hydroxyapatite, particularly onto the newly-formed one.

Chrystallographic studies of model complexes suggest a relationship between the spatial arrangement of O-donor atoms in ligand moieties and the coordination parameters of Ca in hydroxyapatite. Starting from the geminal diphosphonates in MDP (Fig. 13.15.), additional donor groups have been introduced into the central carbon atom in order to increase bone affinity. On this basis hydroxymethylene diphosphonate has been proposed as a better ligand than MDP, DPD is the result of optimization studies with more than 70 ligands. According to Subramanian (1982), in animal studies these new agents show a higher concentration of Tc-complexes only in the normal skeleton and not in lesions. Therefore, search for better agents should be focussed on those with higher lesion/bone ratios.

The use of *radioactive enzyme inhibitors* as potential organimaging agents has been suggested by several investigators. Reversible, irreversible or suicide enzyme inhibitors can be used. Generally structural similarities to substrates are not strictly required since a great variety of structures can be found among inhibitors of an enzyme even without any resemblance to the natural substrate at first sight. Therefore, the "translation" of an inhibitor lead structure into foreign-labelled radiotracer structures may not be the key problem. Limitations arise from the concentration of enzymes and the binding constant. From this point of view, useful target to nontarget ratios are expected for few organs only; the adrenal gland is one of these excep-

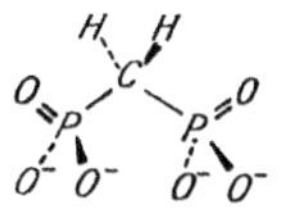

MDP

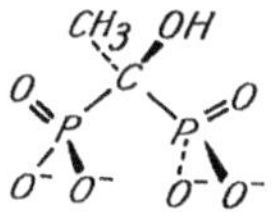

HEDP

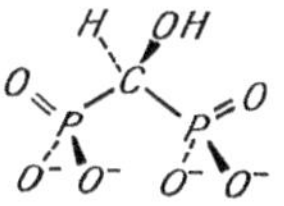

HMDP

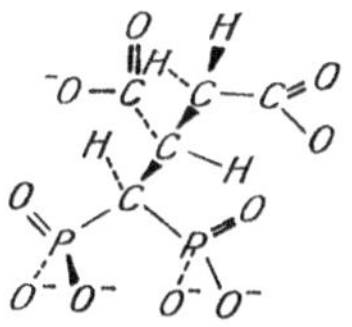

DPD

Fig. 13.15. Comparison of diphosphonate ligand structures

tions as proved by Beierwaltes et al. Burns et al. have dealt with inhibitors of acetylcholin esterase. The high enzyme activity in myocardium has stimulated search for potential myocardial-imaging agents based on the concept of enzyme binding. Ornithine decarboxylase is of interest as a possible enzyme marker of human cancer.

A clases of potential radiopharmaceuticals is based on *receptor binding* /8/. The receptor concept used in pharmacology postulates that the effect of a drug results from interactions of the drug molecule and specific receptor sites occurring either in plasma membrane or with steroids in the cytoplasma of the target cell. Various receptor types and subclasses have been identified and partially isolated, e.g. nicotinic, muscarinic, α- and β-adrenergic, opiate, dopamine, estrogene, etc. receptors. The demonstration of pathological changes in receptor concentration or the appearance of receptors in tumors may be possible. The radiotracer must fulfil various requirements: high affinity to the receptor, low non-receptor binding, very high specific activity and high stability.

10. Immunological concept

For more than two decades it has been the objective to use radiolabelled antitumor antibodies for localization of tumors and metastases as first described by Pressman. By using various animal models only limited success has been achieved by means of antibodies of tumor-associated antigens. Enzymatic cleavage of the immunoglobulin molecule and isolation of the F(ab')2 fragment is one possibility of increasing the target/non-target ratio. Recently there has been new interest in this approach because of the potential improvements in specificity offered by the use of monoclonal antibodies. Further studies, however, are necessary to evaluate the clinical usefulness of this approach.

References

1. *Eckelman, W. C., R. C. Reba:* The classification of radiotracers. J. Nucl. Med. **19** (1978), 1179.
2. *Colombetti, L. G.:* Radiopharmacology (the state of the art). La Ricerva in Clin. Lab. **9** (1979), 281.
3. Principles of radiopharmacology. Vol. I–III. Colombetti, L. G. (ed.). CRC Press, Inc. 1979, Boca Raton USA (Proceedings of the 1st Intern. Symp. on Radiopharmacology, Innsbruck, May 1978).
4. *Johannsen, B.:* Advances in radiochemistry of technetium-99m. 2nd Eur. Symp. on Radiopharmacy and Radiopharmaceuticals, Cambridge 1985.
5. Progress in Radiopharmacology-selected topics. *Cox, P. H.* (ed.). Elsevier/North Holland Biomedical Press, Amsterdam, New York, Oxford 1981 (Proceedings of the 2nd Eur. Symp. on Radiopharmacology, Noordwijkerhout, Netherland, November 1980).
6. *Johannsen, B., H. Spies:* Chemie und Radiopharmakologie von Technetiumkomplexen. 213 p. Zentralinstitut für Kernforschung Rossendorf, 1981.
7. *Feinendegen, L. E., K. Vyska, Chr. Freundlich, A. Höck, H. J. Maculla, G. Kloster, G. Stöcklin:* Non-invasive analysis of metabolic reactions in body tissue, the case of myocardial fatty acids. Eur. J. Nucl. Med. **6** (1981), 191.
8. *Gallagher, B. M., J. S. Fowler, N. I. Gutterson, R. R. McGregor, C.-N. Wan, A. P. Wolf:* Metabolic trapping as a principle of radiopharmaceutical design: some factors responsible for the biodistribution of (^{18}F)2-deoxy-2-fluoro-D-glucose. J. Nucl. Med. **19** (1978), 1154.
9. *Eckelman, W. C., R. C. Reba, R. E. Gibson, W. J. Rzeszotarski, F. Vieras, J. K. Mazaitis, B. Francis:* Receptor-binding radiotracer: a class of potential radiopharmaceuticals. J. Nucl. Med. **20** (1979), 350.

Chapter 14

Basic Principles in Clinical Use of Stable Isotope Pharmaceuticals

Peter Krumbiegel

1. History, classification, definition

Already in 1939, Rudolf Schoenheimer /1/ used stable isotope-labelled amino acids for the first time in medical research. Nowadays the names of Schoenheimer and Hevesy are synonyms for the pioneers of isotope tracer application in physiology and medicine. In spite of the fundamental importance of Schoenheimer's investigations–amongst other things he demonstrated the dynamic balance principle of protein turnover in animals–the use of stable isotopes remained confined to some medical basic research in the following decades. The reason for this was that at that time the availability and the measurement possibilities of stable isotopes were very poor–in contrast of those of certain radioactive nuclides useful for medical purposes. Only the search for increasingly specific methods in physiological functional diagnosis, especially in gastroenterology, stimulated the development of both production and analysis methods for stable isotopes. As a consequence, now a lot of stable isotope-labelled pharmaceuticals ready for application in clinical diagnosis or already in use are known. This will be demonstrated in the 4th section.
Because of the late re-introduction of stable isotopes to pharmacology, there is now some contradiction in using traditional terms such as "**radio**pharmaceuticals", "**radioactive** drugs" or "**radio**pharmacology" in the context with stable isotopes. Therefore, the general term "nuclear pharmaceuticals" should be used connecting both radioactive and stable isotope pharmaceuticals, as given in Figure 14.1.
In analogy to radioactive pharmaceuticals, the definition of stable isotope pharmaceuticals reads:
"Stable isotope pharmaceuticals are pharmaceuticals and substances or preparations of the same kind which are labelled with one or more stable iso-

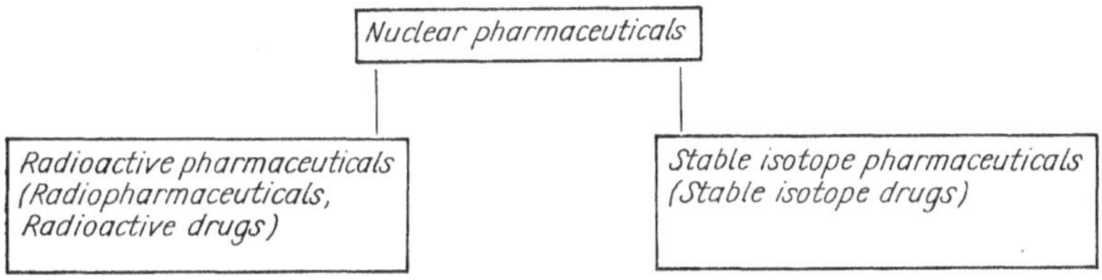

Fig. 14.1. Relation of stable isotope pharmaceuticals to radiopharmaceuticals

topes (that means, which contain at least one chemical element with stable isotopes the relative abundances of which differ from the natural abundances) and which are intended for diagnostic or therapeutic application by use of tracer techniques or isotope effects."
By definition the term "Stable isotope pharmaceutical" is delimited from the term "Stable isotope-labelled pharmaceutical", which does not include the intention of clinical use (but maybe of chemical or pharmacological research).

2. Application field, licence

Klein et al. /2/ proposed three fields of stable isotope application in medicine, in which stable isotopes cannot be replaced by radioactive isotopes:

1. Pediatric and gynaecological studies, in which the risk of radioactive radiation cannot be accepted;
2. Quantification of ultra-low levels of tissue and plasma constituents by inverse isotope dilution assays with an internal standard;
3. Pharmacological studies, particularly with powerful drugs the dosage of which is only a fraction of a milligram, and the maximum possible radioactivity administered hereby does not permit kinetic measurements of drug metabolism.

For clinical practice the field first mentioned seems to be of significance. The other fields require highly sophisticated gas chromatography-mass spectrometry instruments (GC-MS) for isotope measurements which are not common practice in clinics.
On the other hand, there is a further aspect of the need for stable isotopes in nuclear medicine:
For some elements a suitable radioactive isotope as an alternative does not exist. The radioactive nuclides most suitable in nuclear medicine are not identical with the most important biogenic elements. If certain physiological steps must be followed in order to characterize pathological situations, it can be necessary to label one of the elements carbon, nitrogen, hydrogen,

Table 14.1. Radioactive and stable isotopes of some biogenic elements

Elements	Hydrogen		Carbon	
Stable isotopes	^{1}H	^{2}H	^{12}C	^{13}C
(natural relative abundance)	(99.985)	(0.0147)	(98.892)	(1.108)
Radioactive isotopes	^{3}H		^{11}C	^{14}C
(half-life)	12.2 a		20.5 min	5730 a

or oxygen by its own isotope. A foreign label with a typical medical radionuclide would not reflect the real metabolism of the compound under consideration. It is a matter of fact that the isotopes of the biogenic elements mentioned above are either very short-lived cyclotron isotopes or stable isotopes (with the exception of the very long-lived ^{14}C isotope, which is not commonly permitted for clinical use), as seen in Table 14.1. Thus stable isotopes are the only alternative as long as the use of cyclotron isotopes is not suitable.

The absence of any radiation is an essential advantage in the use of stable isotope pharmaceuticals. State control including the restrictions of application known for radiopharmaceuticals does not apply to stable isotope pharmaceuticals. Nevertheless, it had to be investigated, which risk could arise from the what is called isotope effect under certain circumstances. Indeed, in special cases the substitution of ^{2}H (deuterium) for ^{1}H-hydrogen can result in a changed behaviour of the corresponding compound /3/. This was considered in governmental sanctions for the human application of stable isotopes, e.g. in the German Democratic Republic /4/. An important statement of these regulations reads as follows: The application in medicine of chemical substances labelled with the stable isotopes of carbon and nitrogen is licensed without restriction, if the enforced decisions of drug and food legislation are observed.

Such regulations have not yet become common practice in all countries. Therefore, before clinical application of stable isotopes one should inquire about the relevant regulations in one's own country.

3. Availability, measurement principles

As known, a stable isotope is in no case an artificial nuclide but already present in nature as one of the isotopes of the poly-isotopic elements. So the production of stable isotopes for tracer applications means the following: An isotope with a naturally low relative abundance is enriched to a high relative abundance at the expense of the other isotopes of the underlying element. For example, the stable nitrogen isotope ^{15}N with a natural rela-

Elements	Nitrogen		Oxygen		
Stable isotopes	^{14}N	^{15}N	^{16}O	^{17}O	^{18}O
(natural relative abundance)	(99.635)	(0.365)	(99.759)	(0.0374)	(0.2039)
Radioactive isotopes	^{13}N		^{15}O		
(half-life)	10.0 min		2.1 min		

tive abundance of 0.37 at. % (together with 99.63 at. % ^{14}N in natural nitrogen) can be enriched in special chemical plants to 99.9 at. % ^{15}N with a remaining amount of only 0.1 at. % ^{14}N in the nitrogen element. In general, the enriched stable isotope is than introduced into the pharmaceutical in question during its usual chemical synthesis. Stable isotope-labelled compounds are commercially available on a large scale now and can be ordered for tracer technique investigations in vivo, too.
Formulas and names of stable isotope-labelled compounds are written by inserting the isotopes' symbol in square brackets (e.g. $CH_2[^{15}N]H_2COOH$; $[^{15}N]$glycine).
Due to the absence of any radiation signals, stable isotopes or correspondingly labelled compounds must be detected with quite different measurement principles than is done in the case of radionuclides. "Signals" of stable isotopes can be taken on the basis of differences in

- mass,
- optical wave lengths,
- magnetic moments.

The first principle mentioned above is carried out by the wellknown mass spectrometers as the most widely used and most versatile isotope measurement instruments: depending on their mass, ionized and accelerated isotopic molecules are deflected to a certain extent when they pass a magnetic field. The resulting ion beams are collected, detected and displayed.
The second detection principle is used, e.g. in ^{15}N analyzers created and produced first in the German Democratic Republic (models "NOI-5", "NOI-6", "ISONITROMAT"). They are special emission spectrometers with an on-line sample preparation. Such inexpensive instruments have stimulated a wide scale of ^{15}N applications during the last decade.
The third detection principle is implemented with nuclear magnetic resonance (NMR) instruments, and its application has just begun.

4. Clinical applications

In this context, the large field of stable isotope application in pharmacological research cannot be considered. We refer to Baillie /5/ who has reviewed more than 500 articles concerning e.g. the quantification of micro-amounts of drug metabolites of interest, or the follow-up of the metabolism of new drugs. Contrary to that, clinical applications of stable isotope-labelled pharmaceuticals have been confined to relatively few fields up to now, which will be briefly described in the following:

Application of ^{13}C-labelled pharmaceuticals for various "$^{13}CO_2$ breath tests"

The principle is well-known for $^{14}CO_2$ breath tests: A suitable drug is labelled with ^{13}C in such a molecule site (position) which is split off in the physiological situation under consideration and then exhaled as $^{13}CO_2$. From the half-life of $^{13}CO_2$ exhalation, conclusions can be drawn on the functional state of the organ in question. With the substitution of ^{13}C for ^{14}C the corresponding tests became applicable to children and pregnant women. Some of the tests including the test substances are summarized in Table 14.2.

Table 14.2. Application of ^{13}C-labelled pharmaceuticals for $^{13}CO_2$ breath tests

^{13}C-labelled drug	Application field	Authors
[$^{13}C_2$]Aminopyrine	Liver function	Birchner et al. 1975 /6/
[^{13}C]Galactose	Liver function (sugar oxidation)	Shreeve et al. 1976 /7/
[^{13}C]Glycocholate	Ileal function (bile acid deconjugation)	Solomons et al. 1976 /8/
[^{13}C]Trioctanoin	Fat malabsorption	Watkins et al. 1977 /9/
[^{13}C]Lactose	Carbohydrate assimilation	MacLean et al. 1983 /10/
[^{13}C]Methacetin	Liver function (oxidases' activity)	Krumbiegel et al. 1985 /11/

Application of ^{15}N-labelled pharmaceuticals for organ function tests

In analogy to $^{13}CO_2$ breath tests, the turnover rate of a xenobioticum as a measure of an organ function activity can also be estimated with ^{15}N: on the basis of a known metabolism and rate-determining step, the drug is labelled with ^{15}N in such a molecule part and position which is known to be excreted via urine. Then a simple nitrogen and ^{15}N isotope measurement of urine stands for any metabolite identification and quantification. The first test of this kind on the basis of a xenobioticum is conducted with 3 mg [^{15}N]methacetin per kg body mass. The results are comparable to the breath tests results /14/.

A further novel investigation principle is put into research with the introduction of ^{15}N to in-vivo NMR spectroscopy (Table 14.3.).

Table 14.3. Application of ^{15}N-labelled pharmaceuticals for organ function tests

^{15}N-labelled drug	Application field	Authors
^{15}N-Yeast protein	Digestion and malabsorption (coeliac disease)	Crane et al. 1960 /12/
[$^{15}N_2$]Nitrogen	Lung function (residual volumen)	Reinhardt et al. 1984 /13/
[^{15}N]Methacetin	Liver function (oxidases' activity)	Krumbiegel et al. 1985 /14/
[^{15}N]Ammonium chloride	Liver function (urea synthesis capacity)	Jung et al. 1985 /15/
[^{15}N]Ammonium chloride	Hyperammonemia (via blood measurements)	Ukida, 1978 /16/
[^{15}N]Glycine	Liver function (in vivo ^{15}N-NMR spectroscopy)	Buchali et al. 1985 /17/

Application of ^{15}N to label nitrogen-containing endogenous compounds,

to follow protein turnover which can be disturbed owing to a certain disease, to operation and other stresses, to hormone treatment, etc. Such investigations and the corresponding ^{15}N-labelled compounds are summarized in Table 14.4. Mostly the common principle is an end-product analysis with the calculation of whole-body protein turnover based on computer-generated compartmental models.

Table 14.4. Application of ^{15}N-labelled compounds for protein turnover investigations

^{15}N-labelled compound	Application field (protein turnover influenced by:)	Authors
	– Fundamentals –	Waterlow et al. 1978 /18/ Faust et al. 1981 /19/
[^{15}N]Glycine	Operation stress	Hartig et al. 1978 /20/
	Different nourishment of infants	Heine et al. 1983 /21/
	Hormone therapy effects	Willgerodt et al. 1983 /22/
	Top-training of sportsmen	Krumbiegel et al. 1985 /23/
[^{15}N]Ammonium chloride	Hormone therapy effects	Zachmann et al. 1977 /24/
Various D-^{15}N-amino acids	Utilization of D-amino acids	Heine et al. 1983 /21/

Use of deuterated pharmaceuticals

In spite of a lot of proposals to use deuterated pharmaceuticals /25/, only the measurement of total body water by isotope dilution analysis with heavy water seems to be a clinical routine application up to now.
The usage of $^{1}H/^{2}H$ isotope effects is the only known possibility of stable isotope application to alter a *therapeutic* effect of a pharmaceutical.
In Table 14.5. a few examples of proposed deuterated pharmaceuticals are given.

Table 14.5. Some deuterated drugs and their application in medicine

Deuterated drug	Application field	Authors
Indocyanine green in $^{2}H_2O$	Storage stability enhancement	Krumbiegel et al. 1966 /26/
[3'-$^{2}H_2$]Butobarbital	Prolongation of narcotic effect	Tanabe et al. 1969 /27/
$^{2}H_2O$	Body water measurement	Halliday et al. 1977 /28/

As can be seen from the practical examples, the common principle of nearly all clinical applications of stable isotope-labelled pharmaceuticals is a non-invasive procedure: an oral application of the labelled compound and sampling via naturally eliminated breath or urine.

References

1. *Schoenheimer, R., S. Ratner, D. Rittenberg:* Studies in protein metabolism. The metabolic activity of body proteins investigated with L-leucine containing two isotopes. J. Biol. Chem. **128** (1939), 703.
2. *Klein, P. D., P. A. Szczepanik, D. L. Hachey, D. A. Schoeller:* The Argonne Bioanalytical Center: A resource for collaborative biomedical applications of stable isotopes. Proc. Techn. Comm. Meeting Stable Isotopes in the Life Sciences. Febr. 1977, Leipzig, GDR. IAEA, Vienna 1977, p. 313.
3. *Krumbiegel, P.:* Isotopieeffekte. Akademie-Verlag, Verlag Vieweg, Pergamon Press, Berlin—Braunschweig—Oxford 1970.
4. Standard of the GDR: "TGL 14466: Stabilisotop markierte chemische Elemente und Verbindungen. Blatt 3: Anwendung in der Humanmedizin und Lebensmittelindustrie"; valid since 1. 1. 1970.
5. *Baillie, T. A.:* The use of stable isotopes in pharmacological research. Pharmacol. Rev. **33** (1981), 81.
6. *Birchner, J., A. Küpfer, R. Platzer, J. Gikalow, R. Preisig:* Breath analysis of hepatic microsomal function in man. Proc. 2nd Int. Conf. Stable Isotopes Oak Brook 1975, 252.
7. *Schreeve, W. W., J. D. Shoop, D. G. Ott* et al.: Test for alcoholic cirrhosis by conversion of ^{14}C- or ^{13}C-galactose to expired CO_2. Gastroenterology **71** (1976), 98.
8. *Solomons, N., D. Schoeller, D. A. Klein, J. Wagonfeld, D. Ott, F. Viteri, J. Rosenberg:* Diagnosis of bacterial overgrowth or ileal dysfunction in children by use of a stable isotope breath test, ^{13}C-glycocholate. Clin. Res. **24** (1976), 291 A.
9. *Watkins, J. B., D. A. Schoeller, P. D. Klein, D. G. Ott, A. D. Newcomer, A. F. Hofmann:* ^{13}C-trioctanoin: a nonradioactive breath test to detect fat malabsorption. J. Lab. Clin. Med. 90 (1977), 422.
10. *MacLean, W. C.,* jr., *B. B. Fink, D. A. Schoeller, W. Wong, P. D. Klein:* Lactose assimilation by full-term infants, relation of carbon-13 and hydrogen breath test with feceal carbon-13 excretion. Pediatr. Res. **17** (1983), 629.
11. *Krumbiegel, P., K. Günther, H. Faust, G. Möbius, K. Hirschberg, G. Schneider:* Nuclear medicine liver function tests for pregnant women and children. 1. Breath tests with [^{14}C]methacetin and [^{13}C]methacetin. Eur. J. Nucl. Med. **10** (1985), 129.
12. *Crane, C. W., A. Neuberger:* Absorption and elimination of ^{15}N after administration of isotopically labelled yeast protein and yeast protein hydrolysate to adult patients with coeliac disease. I. Rate of absorption of ^{15}N-yeast protein and yeast protein hydrolysate. Brit. med. J. (London) **2** (1960), 815.
13. *Reinhardt, R., J. Schauer, H. Faust:* Eine neue isotopentechnische Methode zur Bestimmung des Residualvolumens. Z. Erkrank. Atm.-Org. **162** (1984), 128.
14. *Krumbiegel, P., K. Hirschberg, H. Faust, K. Günther, G. Schneider:* Nuclear medicine liver function tests for pregnant women and children. 2. A new test method via urine using [^{15}N]methacetin. Eur. J. Nucl. Med. **11** (1985), 58.

15. *Jung, K., K. Hirschberg, H. Faust, R. Matkowitz:* A liver function test with ^{15}N-labelled ammonium chloride. Eur. J. Nucl. Med. **11** (1985), 62.
16. *Ukida, M.:* A clinical application of the measurement of ^{15}N-ammonia using chemical ionization mass spectrometry: ^{15}N-ammonium chloride tolerance test in liver cirrhosis patients with hyperammonemia. Koenshu–Iyo Masu Kenkyukai **3** (1978), 121.
17. *Buchali, K., W. Gründer, P. Krumbiegel, H.-J. Blesin, H. Faust:* Erste Erfahrungen mit der ^{15}N-NMR-Spektroskopie an Rattenorganen. Radiobiol. Radiotherap. **26** (1985), 543.
18. *Waterlow, J. C., P. J. Garlick, D. J. Millward:* Protein Turnover in Mammalian Tissues and in the Whole Body. North Holland Publ. Comp., Amsterdam 1978.
19. *Faust, H., H. Bornhak, K. Hirschberg, K. Jung, P. Junghans, P. Krumbiegel, R. Reinhardt:* Klinisch-chemische und isotopenanalytische Methoden zur Untersuchung des Stickstoff-Stoffwechsels mit ^{15}N beim Menschen. Methodenkatalog. ZfI-Mitt. **36** des Zentralinstituts für Isotopen- und Strahlenforschung der Akademie der Wissenschaften der DDR, Leipzig 1981.
20. *Hartig, W., H.-D. Czarnetzki, H. Faust, K. Hirschberg, K. Jung, P. Junghans, K. Wetzel, H. Willgerodt:* Die Bedeutung des stabilen Stickstoffisotops ^{15}N für die humanmedizinische Forschung und Praxis. DDR-Medizin-Report **7** (1978), 563.
21. *Heine, W., I. Richter, Chr. Plath, K. Wutzke, H.-J. Stolpe, M. Thieß, J. Töwe:* Die ^{15}N-Tracertechnik in der pädiatrischen Forschung. DDR-Medizin-Report **12** (1983), 833.
22. *Willgerodt, H., W. Hoepffner, R. Reinhardt, P. Junghans, E. Keller:* Verbesserung der Differentialdiagnostik des Pseudohermaphroditismus masculinus durch den Einsatz von ^{15}N-markierten Aminosäuren. Methodische Aspekte. ^{15}N-Symposium Rostock 1983. ZfI-Mitt. **77** des Zentralinstituts für Isotopen- und Strahlenforschung der Akademie der Wissenschaften der DDR, Leipzig 1983, p. 254.
23. *Krumbiegel, P., K. Kühne, H. Faust, H. Bornhak, P. Junghans, H. Zerbes:* Zur Ermittlung von Parametern des Proteinumsatzes mit Hilfe der ^{15}N-Tracertechnik bei Sportlern. Medizin und Sport **25** (1985), 13.
24. *Zachmann, M.* et al.: Influence of testosterone on urinary ^{15}N balance in normal subjects and patients with testicular feminization. Clin. Chim. Acta **77** (1977), 147.
25. *Blake, M. I., H. L. Crespi, J. J. Katz:* Studies with deuterated drugs. J. Pharm. Sci. **64** (1975), 367.
26. *Krumbiegel, P., O. Serfas:* Die Stabilisierung von Indocyaningrün in Injektionslösungen mit Hilfe von schwerem Wasser und Aminosäuren. Arch. Pharmazie **299** (1966), 432.
27. *Tanabe, M., D. Yasuda, S. LeValley, C. Mitoma:* The pharmacologic effect of deuterium substitution on 5-n-butyl-5-ethylbarbituric acid. Life Sci. **8** (1969), 1123.
28. *Halliday, D., A. G. Miller:* Precise measurement of total body water using trace quantities of deuterium oxide. Biomed. Mass Spectr. **4** (1977), 82.

Chapter 15

Requirements for the Development and Evaluation of New Radiopharmaceuticals

Gertrude Pfeiffer

1. Introduction

Radiopharmaceuticals or radioactive drugs are substances which are labelled with radioisotopes intended for human use. Therefore, the criteria for safety and efficacy have to include aspects of radiation protection and pharmacy as well. In most countries using radiopharmaceuticals there are strict governmental regulations concerning the handling and use of radioactive material. They are aimed at keeping the radiation hazard to patient and personnel as low as possible. On the other hand, regulations for production and administration of pharmaceuticals, too (WHO-regulations for Good Manufacturing Practice, etc.). The chemical and physical properties of the compounds, their "specifications", methods for quality control and the requirements for purity are stated in the different pharmacopoeas. But as far as radiopharmaceuticals are concerned, there is a lack of regulations and the pharmacopoeas are never up-to-date. The same problem concerns the registration procedures: they are under discussion in nearly all countries, but there are only few of them with well-defined working procedures. Therefore, till now the specifications and the limits of the highest tolerable impurities in radiopharmaceuticals are determined by the manufacturers themselves, and it is their responsibility to prepare them as savely and efficiently as possible and to guarantee a high and constant level of quality. It is obvious that the responsibility of the manufacturer ends with the delivery to the hospital. All further manipulations, including dispensation, and, last but not least, the labelling of non-radioactive compounds with short-lived radionuclides (e.g. technetium-99m) belong to the responsibility of the user /1/.

2. Development of new radiopharmaceuticals

When designing a new radiopharmaceutical it must first of all be decided, if the compound is intended for use in diagnosis or in therapy. The ideal radiopharmaceutical for diagnostic procedures must be different from the ideal substance for therapy concerning the physical properties of the radionuclide.

A pure gamma emitter will be optimal for labelling a diagnostic compound; for therapy, particle emitters are desirable, because these radionuclides combine a high radiation dose for the patient with a low dose for the personnel.

2.1. The radionuclide

Some physical properties of radionuclides which are essential for their optimal use for diagnostic or therapeutic purposes are listed below (Table 15.1.).

The range of optimal gamma energy depends on the equipment used and the process studied. Gamma rays of low energy show a high self-absorption in the body and correction methods for quantification are required. The errors will be higher, if the organ studied is large and situated in a non-defined distance from the body surface. High-energetic gamma rays, however, will penetrate the septa of the collimators, and decrease the resolution. Today, using the general instrumentation (Gamma cameras, type Anger) at the department of nuclear medicine, the optimal energy range for detecting gamma rays from inside the body lies between 100–250 keV. For positron emitters the positron camera will be the optimal equipment, but these cameras are rare today and limited to some special institutions.

The optimal half-life of radionuclides used for diagnostic procedures should be in a reasonable relation to the duration of the study. Wagner pointed out in his book /2/ that the duration of the study (T), multiplied by a factor of 0.69, will result in the optimal half-life for any special study. A longer half-life than necessary would increase radiation to the patient unnecessarily and should therefore be avoided. For short studies radionuclides with a very short half-life seem to be optimal due to the radiation dose to the patient, but there will be an increase of dose to the staff because they deal with very high activities. The loss of activity through decay during preparation, purification, packing and transportation of a fluorinated (^{18}F: half-life 110 min) compound can easily be in the range of 50 or more.

2.2. The radiopharmaceutical

Considering the design of a new radiopharmaceutical or selecting the optimal compound for a special diagnostic investigation zum Winkel /3/ listed the following criteria:

- high detectability combined with a low radiation dose to the patient,
- high specific activity,
- identity,
- simple preparation/availability,
- sterility and apyrogenicity.

Table 15.1. Some physical properties of radionuclides for diagnosis and therapy

	Diagnosis	Therapy
Mode of decay	γ, EC, IT	α, β
% Efficiency	high	high
Energy	100–250 keV	depending on lesion
Half-life	$0.69 \times T$	depending on lesion

Detectability

This point deals with the physical properties of the radionuclide used and was discussed above (2.1.). The optimal radionuclide for diagnostic procedures is defined by the following properties: pure gamma emitter with a high photon efficiency in the energy range between 100–250 keV and a half-life of $0.69 \times T$ (investigation time). In daily practice the optimal radionuclide for localization and short function studies will therefore be the well-known technetium-99m (half-life 6 h, only gammas of 140 keV).

Specific activity

Labelled compounds are mainly used as "tracers". They should not show any pharmacological effects. Therefore the specific activity (activity/weight) should be high. But there are some limitations thanks to different reasons to reach a very high specific activity of radiopharmaceuticals:

a) maximal theoretical specific activity of the radionuclide: it can be calculated by transforming the following equation:
Mass of 1 Ci (g) $= 0.88 \times 10^{-13} \cdot A \cdot t/2$
(A: atomic mass, t/2: half-life in s)
Max. spec. activity (Ci/g) $= 1 : 0.88 \times 10^{-13} \cdot t/2$
Some examples demonstrate the influence of half-life on the maximal theoretical specific activity.

Technetium-99m half-life	6 h max.	5.3 Ci/μg
Iodine-123	13 h	2 Ci/μg
Iodine-131	8 d	123 mCi/μg
Iodine-125	60 d	15 mCi/μg
Carbon-14	5,730 a	4 μCi/μg

b) maximally obtainable specific activity: in practice the theoretical value cannot often be obtained because of the production process: the production of a radionuclide in a reactor using the (n,γ)reaction (bombardment with thermal neutrons) result in radionuclides of low specific activity because the new nuclide is an isotope of the target nuclide (e.g. ^{23}Na (a,γ) ^{24}Na). Pure radionuclides are obtained by using accelerators or generator systems.

c) labelling technique: some labelling reactions, like exchange labelling, yield radiopharmaceuticals with lower specific activity than direct labelling. The labelling efficiency itself plays an important role for the specific activity of the end-product, too.

There is one more important point in practice. The handling of radioactive material of high specific activity (called "carrier-free", or rather "no-carrier-added"), is very difficult: unspecific adsorption reactions occur on any surface, so that the loss of radioactive material can be very high. Very often, therefore, a carrier had to be added in order not to lose the whole substance. The main advantage of radiopharmaceuticals with high specific activity is its use as tracers: no pharmacological effects, no influence on the process studied and even toxic substances can be used because of the very low amount administered to the patient. Examples are fluorinated sugars, or thallium. The main disadvantages are the problems connected with the handling mentioned above and the risk of radiation damage during preparation and storage due to the high dose rate in a small mass and/or volume. Some other pros and cons are listed below:

advantages:
- tracer: no pharmacological effects,
- no problems with toxicity,
- even "unsoluble" material remains in solution;

disadvantages:
- high risk of radiation damage of the molecule,
- loss through unspecific adsorption,
- reactions do not follow the chemical laws for equilibrium or reactions kinetics because of the small mass.

Identity

The requirement of identity means that the labelled compound, the tracer, should be identical with the physiological substance or the drug, whose behaviour in the body is measured. There are only very few substances that will completely fulfil this requirement; the big rest will show different degrees of similarity.

a) chemical identity: this ideal case as rare, restricted to simple ions (Na, K, Fe, I), labelled iodinated contrast agents, vitamin B-12, thyroxin, and organic compounds labelled with radioisotopes of hydrogen, carbon, nitrogen or oxygen.
b) structural analogues: one atom in a molecule will be replaced by the radioisotope of another nuclide which shows very high structural similarity (e.g. ionic radius). Examples are the replacement of calcium by strontium-85, the replacement of sulphur by selene-75 in methionine and the functional replacement of potassium by thallium-201 in the muscle cells of the heart.
c) biological analogues: some hormones and drugs may be labelled with heterogeneous radionuclides without changing their physiological and biological behaviour. High molecular compounds, like proteins, can sometimes be labelled very easily with radioiodine at a site of the molecule not responsible for their biological or immunological properties. The labelling of smaller molecules, like steroids, is much more difficult. A special knowledge is necessary to find out the critical and the non-critical sites of the molecule to be labelled, where the radionuclide can be introduced without reducing efficiency.
d) functional analogues: for function studies of some organs or for localization studies radiopharmaceuticals can be used which only show a distinct accumulation in the target organ or a mode of distinct excretion. Most of the technetium-radiopharmaceuticals belong to this group (Tc-DTPA is excreted by glomerular excretion, Tc-IDAs are accumulated in the liver and excretion via bile in the intestinum).
e) particle size: only due to the physical property of the size, labelled particles of colloids are selected for studies of the reticuloendothelial system or for flow studies (liver and lung scintigraphy).

Simple preparation/availability

Radiopharmaceuticals with half-lives of more than one day are mainly available from stock.

Half-lives of less than one day give rise to a lot of logistic problems concerning production, quality control, transportation and delivery to the hospital These disadvantages can be avoided if the radionuclide can be obtained from a generator system (typical example: the technetium generator). Using the radionuclides, eluted from the generator, different radiopharmaceuticals can be prepared, if the necessary non-radioactive chemicals are provided and the labelling process is simple and safe. Some requirements for these labelling procedures ("kit labelling"):

- simple, one-step preparation: adding the eluted radionuclide to a provided sterile preparation of the pharmaceutical,
- quick because of the short half-life of the nuclide,
- only one well-defined end-product, no side reactions,
- high labelling efficiency to avoid time-consuming and risky purification steps,
- mild labelling conditions: neither extreme pH-values (optimal around the neutral range) nor high temperatures.

Sterility, apyrogenicity

Radiopharmaceuticals are generally injected intravenously; therefore, the requirements of the local pharmacopoeas concerning drugs for parenteral use must be met. The only exception is that the radiopharmaceuticals can be administered before the result of the sterility test is known. As to the same volume of injected material, pyrogentests are only abligatory, if the compound will be injected in the subarachnoidal space.

3. Evaluation of new radiopharmaceuticals

The evaluation process for any drug or substance for human use includes tests for safety and efficacy. The basis is for both parameters is the "specifications", which are defined and constant over a longer period. To guarantee these specifications, the manufacturers built up a quality assurance system. It contains:

- control of materials supply,
- qualitative and quantitative control of raw materials,
- prescriptions of target preparation and bombarding procedures,
- prescription for production of the radiopharmaceutical,
- quality control of intermediates and end-product,
- dispensation control,
- packing, labelling, package insert,
- testing equipment,
- inspections.

As far as the user is concerned, the most important parameters are the quality control of the end-product and clinical trials, which guarantee safety and efficacy of the product.

3.1. Analytical quality control and safety tests

Reviews of methods for analytical quality control and their application to special radiopharmaceuticals have been published /4, 5/.

Determination of radioactivity

Before application to the patient, the activity of the radiopharmaceutical must be determined by the user. This work must be done in addition to the data stated on the label, because the user is responsible for the dose administered. The device used for measurement must have been once calibrated and checked on:

- influence of volume,
- influence of geometry (bottles/syringes),
- influence of material (glass/plastic),
- function: using a calibrated standard source, the equipment must be tested in regular intervals.

Radionuclidic purity

The origin of systematic impurities in radionuclides are impurities in the target material or side-reactions during the production process. They are quite frequent, if the radionuclide is produced in accelerators. They cannot be avoided completely, only reduced by choosing optimal bombarding conditions. For example, iodine-123 is mainly produced by two different reactions: in both of them impurities will occur at a very small amount, either ^{124}I or ^{125}I. Non-systematic impurities are accidental contaminations or the breakthrough of the mother nuclide in a generator system.
Methods for the determination of radionuclidic purity can be very simple, like the determination of molybdene-99 in technetium-99m, because these two radionuclides emit gamma rays of very different energies. A computerized Germanium-Lithium detector system should be used more frequently, because of the very small amounts of the impurities. If impurity has a longer half-life than the main radionuclide, or if it decays by particle emission, an additional unnecessary radiation dose to the patient will be the result.

Radiochemical purity

Radiochemical impurities may be dangerous for two reasons: if the impurity has a longer biological half-life than the radiopharmaceutical, the radiation dose to the patient will increase (e.g. radioiodide in labelled hippuric acid). The different biodistribution of impurity can cause misinterpretations of registered data and lead to erroneous diagnosis.
For the determination of radiochemical purity mainly chromatographic methods are used. Nearly all procedures published are listed in the German STH-reports /6/.

Chemical purity

Some substances, even in tracer amounts (metals from the target material or chemicals), may disturb the labelling procedures or change the body distribution. Therefore, limits must be fixed and analytical tests performed.

Pharmaceutical purity
As pointed out in Chapter 2.2., radiopharmaceuticals have to meet requirements for parenteral drugs.

Chemical toxicity
Using radiopharmaceuticals, chemical toxicity is not a big problem, because the administered amount is very small (within the range from pg to mg), and they are only applied once, so that no chronic toxicity effects may occur. Of cause, it is necessary to undertake some toxicity tests in the development of a new radiopharmaceutical; the requirements are mostly prescribed by the local authorities.

Radiation toxicity
The determination of radionuclidic and radiochemical purity guarantie that the radiation dose to the patient will be as low as possible during diagnostic procedure. The regulations for handling radioactive material will protect the staff from radiation, if they are observed. In therapy the purpose of the administration of the radiopharmaceutical is a localized radiation: for these compounds it is essential that the accumulation in the lesion is very high. To estimate the radiation dose, the biodistribution and the pharmacokinetics of the radiopharmaceutical must be determined.

3.2. Efficacy tests

Biological tests
Developing a radiopharmaceutical for diagnostic studies, biodistribution in animals shows, whether the accumulation in the target organ will be high enough to carry out the desired investigation. For localization studies an uptake of nearly 100 % would be optimal. If the intended use is function studies, pharmacokinetic data must be collected. These tests are conducted by using animals; mainly two different species are used. The data of biodistribution and excretion rates allow a first estimation of the radiation dose.

Clinical trials
Last but not least, a clinical study must prove the safety and the efficacy of the new radiopharmaceutical as well. Carrying on these studies local requirements for "clinical tests" must be met.
Clinical tests must include the following points:
- investigations into normals/volunteers:
  - body distribution, excretion mode and kinetics,
  - estimation of radiation dose,
  - registration of adverse reactions,
  - accumulation in the target organ;
- investigation into patients:
  - suitability test with selected patients,
  - pharmacokinetics in patients with disease studies,
  - abnormal biodistribution in patients with abnormal kidney function,
  - estimation of radiation dose under pathological conditions,
  - studies into patients with unknown disease to verify suitability.

4. Summary

The safety and efficacy of a new radiopharmaceutical for diagnostic or therapeutic use must be proven very carefully before it can be administered to patients. If there are any governmental regulations on production, quality control and registration of radiopharmaceutical, they have to be observed. The main criteria for safety and efficacy can be summarized:

- high quality of the product: high and constant purity, specifications' guarantee,
- preclinical studies: the determination of biodistribution and metabolism for estimating the radiation dose and checking suitability,
- clinical trials on normal volunteers, selected patients to verify the efficacy of the radiopharmaceutical.

References

1. *Kristensen, K.:* Preparation and control of radiopharmaceuticals in hospital. IAEA Technical Report Series, No. 194, 1979.
2. *Wagner, H. N.:* Principles of Nuclear Medicine. WB Saunders, Philadelphia 1968.
3. *zum Winkel, K.:* Nuklearmedizin. Springer, Berlin 1975.
4. *Pfeiffer, G.:* Analytical quality control. Eur. Symp. on Radiopharmacy and Radiopharmaceuticals (ed.): K. Kristensen, 1982, p. 66.
5. *Phan The Tran, R. Wasnich:* Practical nuclear pharmacy. Banyan Enterprises, Honolulu 1979.
6. *Heide, L.*, et al.: Analytik radiochemischer Verunreinigungen in Radiopharmazeutika. Institut für Strahlenhygiene des Bundesgesundheitsamts, STH-Berichte 15, 1979/7, 1980/9, 1980/5, 1981.

Chapter 16

Good Radiopharmacy Practice

Knud Kristensen

1. Introduction

High quality of radiopharmaceuticals is of paramount importance for nuclear medicine. High quality means a uniform quality concerning an agreed standard. Radiopharmaceuticals should have uniform properties, be effective with regard to the intended use and have a minimal risk compared to the benefit obtained /1/. The quality of a radiopharmaceutical is not just determined by the production methods or the quality control carried out. It may be affected also by the handling and storage, and a quality assurance programme for the whole system must be established /2/. The quality of a radiopharmaceutical is defined by setting a standard for the product and the manufacturing process. Good Radiopharmacy Practice /3/ is defined as a previously defined manufacturing process carried out and recorded by a trained and qualified staff provided with the necessary facilities including adequate premises, suitable equipment, correct material and approved procedures. This Chapter will deal with the handling of radiopharmaceuticals at hospitals. We shall first discuss some general principles of Good Radiopharmacy Practice by focussing on individual elements and secondly we shall discuss the practical handling of the different types of radiopharmaceuticals at hospitals.

2. Elements of Good Radiopharmacy Practice

2.1. Personnel

The availability of a sufficiently well-trained and qualified staff is one of the most important elements of Good Radiopharmacy Practice. The duties and responsibilities of each individual should be clearly defined and described in writing. In a system like a nuclear medical department it is also important that qualified personnel can take over responsibilities whenever people are on holiday or on sick leave. The individual nuclear medical department may not be able to have experts in every field, but should clearly be able to draw on expertise from outside in such areas. People involved in the production and quality assurance of radiopharmaceuticals should have had practical experience and theoretical training in areas such as pharmacy (including microbiology), chemistry (including radiochemistry), radiation protection, radiopharmacology and radiopharmacy. It is important that profound training is provided in principle by Good Manufacturing Practice, hygiene and radiation hygiene. Additional training, in order to keep the staff up-to-date in all relevant areas, must be provided continuously. In particular hygiene and radiation protection aspects should be dealt with at least every six months.

2.2. Premises

The design of rooms for the preparation of radiopharmaceuticals has two main purposes: to protect the operator from radioactivity, and to protect the product from contamination by the operator and the rest of the environment. This is achieved by a design where cleaning is facilitated by the construction and type of furnishing used, where the air is continuously exchanged with clean air and where a system of barriers is built up around the product to prevent very large differences in contamination levels. Such barriers may, for example, consist of a closed vial, a contained work station with laminar air-flow, and a ventilated laboratory room within a building. A source of contamination is brought in whenever the operator or his hands enter the system. The risk is minimized by dressing the operator in suitable protective clothing, which will also serve to protect him from the product. The ventilation system also serves to prevent the spread of radioactivity. The design should take account of any special environment conditions such as tropical climate. With most radiopharmaceuticals, where operations are very simple and nonautomated, the greatest risk of contamination of the system is produced by the operator rather than by the rest of the environment. In general, hygienic and radiation protection requirements for the design of laboratories are of the same nature. Both hygienic and radiation protection requirements are the same for the furnishing of rooms: easily cleanable surfaces with no sharp corners and free from cracks and open joints. As far as ventilation systems and the installation of sinks are concerned, there may be opposite considerations. In areas designed for handling radioactivity a negative pressure in relation to the environment is normally established in order to avoid releases of activity. A tap with a sink is essential for flushing away any accidentally spilled activity from the operator. An aseptic area for the preparation of radiopharmaceuticals is kept at positive pressure to keep out particles and microorganisms. A sink and the connected sewer is a possible source of bacterial contamination.
If the amount of radioactivity handled exceeds certain limits, radiation protection may require special design features. If sterile radiopharmaceuticals are prepared, special features are also needed.
Premises must be of sufficient size, so that there is ample room for different working procedures. This also facilitates good housekeeping and cleaning. The same applies to the space needed for paper work, storage, washing glassware, etc.

2.3. Equipment

It is good practice that all equipment be kept well-maintained, calibrated and controlled for proper function by frequent checks. This particularly applies to activity meters (dose calibrators), monitoring equipment, autoclaves and sterilizers. Laminar air-flow units should be tested at frequent intervals.

2.4. Documentation

Documentation is absolutely indispensable in a quality assurance system. The main purposes are: to define and describe the system of control to reduce the risk of error that purely verbal communication introduces and to make it possible to trace an individual product back through the whole process including the utilization and disposal of starting materials, packaging materials and finished products. Use of batch numbers permits easy logical filing, thus enabling rapid retrieval.

All documentation material must be kept up-to-date and instructions, master formulas, etc. must be written in such a way that no doubt can arise about which issue is valid and which changes have been made. All materials should be dated and should be authorized by the relevant competent persons. All out-dated material must be collected. The documentation system should be regularly reviewed and revised accordingly. It may consist of a system of general procedure documents and documents related to each individual product. The documentation system should be designed to include both pharmaceutical and radiation protection aspects.

2.5. Working methods

Safety and efficacy of the radiopharmaceutical and safety of the personnel is best achieved by the use of well-defined working methods. Special methods must be used for sterile products, particularly when the product cannot be terminally sterilized. The specification for sterile products (max. 1 un-sterile per 10^6) obviously cannot be found by sterility testing. It is necessary to avoid contamination with microorganisms during the preparation procedure.

2.6. Quality control

A quality control system must be established for each radiopharmaceutical prepared. It may consist of complete analytical testing of each individual batch or may rely on control of the method by continuous surveillance of certain parameters independent of the release for use of individual batches. Such a system will always be supplemented by a few checks on the individual batch .It is fully acceptable in a well-established production system to deliver radiopharmaceuticals to the user before the results of the quality control test are available. Some tests may then be finished before the product actually reaches the user, others such as the sterility test can only be used as a control of methods. The acceptance of such a philosophy stresses the importance of a well-functioning recall system which makes it possible to get the product back from the user.

Quality control analyses are of equal importance for raw materials, packing materials and finished products, and they are dependent on suitable facilities with good equipment and a well-trained staff. In particular testing for microbial contamination and testing using animals have to be performed at laboratories well separated from production facilities but also for other laboratories; cross-contamination between the laboratory and production areas should be avoided.

Quality control performed as control of methods was first introduced in connection with sterility and pyrogen testing where, because of the difference between the half-life of the radiopharmaceutical and the time needed for the test to be performed, it was absolutely necessary to accept this principle. The very limited stability of many radiopharmaceuticals means that products of a lower quality may result from the fact that they have to be held back until quality control results are available. With the well-documented production system now established it is an obvious advantage to direct more effort into continuous surveillance of the method than into acual batch testing.

3. Handling of radiopharmaceuticals at hospitals

3.1. General aspects

Most radiopharmaceutical handling procedures at hospitals are very simple, such as dispensing patient doses from stock solutions (handling of ready-for-use radiopharmaceuticals) or simple preparations of short-lived radiopharmaceuticals from radionuclide generators and preparation kits, in particular preparation of sodium pertechnetate (Tc-99m) injections and preparation of Tc-99m-labelled radiopharmaceuticals. In principle GRP as described is used, but there are reasons for a simpler system, as a greater part of the work has been done before the radiopharmaceuticals or semi-finished products arrive at the hospital. The procedures that have to be applied are very simple, normally involving only one or two transfers between closed containers, and raw materials are supplied as ready-for use stock solutions or as semi-finished sterile products, where quality control of raw materials and most aspects of quality control of the final product have already been dealt with /4–8/.

The half-lifes of Tc-99m and In-113m of 6 hours and 100 minutes mean that labelled products with these radionuclides can only be used within the same working day they are prepared. They are normally applied intravenously within a very few hours and may be stored at 2–5 °C between preparation and injection; that is why problems of microbiological safety are much less dominant than with other sterile pharmaceuticals.

3.2. Ready-for-use radiopharmaceuticals

Such radiopharmaceuticals will very often require dispensing of patient doses because of the short half-life of the radionuclide whereby the radioactive concentration will change continuously. The laboratory facilities required for this very simple procedure may be rather limited when dispensing is carried out at an individual hospital for immediate use.

Before a radiopharmaceutical is ordered, a decision must be made as to what specifications have to fulfilled. Reference may be made to a pharmacopoeia monograph in order to specify a distinct quality in an unequivocal manner. It is important that control is exercised and that the product then received is described in the same way.

Requirements for storage of the radiopharmaceutical must be followed both with regard to temperature and radiation protection.

No quality control will normally have to be performed with ready-for-use products. This has been observed by the manufacturer. Visual inspection, control of paperwork and control of the activity of individual patient doses is all that will be needed.

3.3. Preparation of sodium pertechnetate (Tc-99m) injections from Tc-99m generators

The starting material is a Tc-99m generator produced so that the eluted product will fulfil the requirements of a pharmacopoeia or a similar specification. The generator is delivered with all necessary accessories for elution including the sodium chloride solution. No other solutions should be used.

The radiation protection aspects and hygienic conditions must be considered when locating a Tc-99m generator. Extra radiation shielding is very often needed to meet radiation hygiene rules. Tc-generators should be placed in a room with a proper hygienic standard. As there is no significant risk of airborne radioactive contamination, no fume hood is needed. If an isotope laboratory with a very good hygienic standard can be set aside for the generator operation alone, this may be acceptable; but if the room is very often used, the generator should at least be protected by a laminar air-flow cabinet.

Another special problem is autoclaving or membrane filtration of the final product. The pertechnetate solution will stand both treatments, and normal pharmaceutical practice would indicate the use of autoclaving as a final step. In this particular case an increased radiation dose risk may follow, and in a well-organized and controlled system it is considered to be acceptable to omit this procedure.

Quality control should include properties that may be affected by transportation of the generator and operation at the hospital: elution yield, Mo-99 in eluate and control of aseptic technique.

3.4. Tc-99m-labelled radiopharmaceutical prepared from preparation kits

Preparation is carried out using sodium pertechnetate (Tc-99m) injection and preparation kits. It is assumed that such kits are produced in accordance with generally good pharmaceutical manufacturing practice and according to fixed specfications either in pharmacopoeias or in registration documentation.

As far as Tc-99m prepared from generators is concerned, responsibility is shared between the manufacturer of the preparation kit and the person responsible for preparing the Tc-99m-labelled radiopharmaceutical at the hospital.

Generators and kits may originate from different manufacturers, as not all kit manufacturers produce generators and not all generator manufacturers produce all types of kits. Thus, there may be the problem of compatibility between generators and kits. Information should be requested from the manufacturer but the final responsibility must, unfortunately, rest with the hospital, unless kit and generators from the same manufacturer are used.

All kit procedures are performed in closed systems and the resulting products are used within one working day. Most kits require only a single-step procedure. The requirements for hygienic conditions and radiation protection are therefore similar to those required for dispensing ready-for-use radiopharmaceuticals. Manufacturers give full instruction on preparation details. They should be strictly followed and interpreted to the technical staff performing the daily procedures /8, 9/.

Analytical quality control of routine preparation when using Tc-99m kits and generators should not be needed. It is, however, useful to be able to perform a simple radiochemical purity test for use in trouble-shooting whenever strange imaging results are obtained.

In order to be able to use such a test, it has to be practised. It is therefore recommended that is should be used whenever a new type of Tc-99m generator is introduced or a new technician is acquainted with his new work, and possibly at regular intervals.

3.5. Labelling of red blood cells and other biological samples with the help of preparation kits

Labelling of samples from patients with radionuclides presents the same problems as the preparation of other radiopharmaceuticals, and the same principles as already described should be used.

There are two complicating factors:

a) Procedures are often multi-steps and may include centrifugation;
b) Patient samples may contain infectious material, such as hepatitis virus, presenting an added risk to the operator and the risk of cross-contamination between samples.

Although preparation takes place in closed systems, the added infection risk requires that the procedure be separated from other radiopharmaceutical

procedures. The products are growth media for microorganisms and the requirement of a good aseptic technique is therefore stressed.
As far as possible, work should take place in a downward-flow laminar airflow (LAF) work station situated in a separate room. If work has to be carried out in the same room as normal radiopharmaceutical preparation work in a closed system, it is considered necessary to have a separate LAF unit or a suitable box structure for handling these biological materials.
The simplified quality control testing may have to include, apart from the problem of labelling yield, consideration of possible damage to the biological sample. Control of the aseptic technique must be performed.

3.6. Other radiopharmaceuticals

In-113m radiopharmaceuticals can be prepared according to the same principles as for Tc-99m radiopharmaceuticals which means in a rather simple system.
When the hospital starts production by preparing its own kits or by labelling biological material such as monoclonal antibodies by complicated procedures such as iodination, the problems become much more complex.
With such "home-made" radiopharmaceuticals the hospital must guarantee all aspects of product quality, including:
- radionuclide purity,
- specific activity,
- radioactive concentration,
- pH,
- isotonicity,
- particle size (if relevant),
- sterility, and
- apyrogenicity.

4. Surveillance of the total system

The quality of the radiopharmaceutical, after administration to the patient, can be observed by physicians during and after the nuclear medical procedure. During the interpretation of images the physician should be aware of false biological distributions.
Radiopharmaceutical defects detected by quality control procedures or by similar observations should be reported to the manufacturer as well as to any relevant national, regional, or international committee which may have been set up to monitor and record such defects /10–12/.
Unexpected adverse reactions, such as pyrogenic, vasovagal and allergic ones, should also be reported to the manufacturer and to any relevant national, regional, or international committee. Occurrence of these events must be fully documented.
Such reporting systems are considered to be of primary importance in order to obtain an early detection of any defect or adverse reaction.

References

1. *Kristensen, K., E. Nørbygaard:* Safety and Efficacy of Radiopharmaceuticals, Martinus Nijhoff, The Hague, 1984.
2. Quality Assurance in Nuclear Medicine, WHO, Geneva 1982.
3. *Kristensen, K.:* Preparation and Control of Radiopharmaceuticals in Hospitals. IAEA, Tech. Rep. Ser. 194, Vienna 1979.
4. British Institute of Radiology. Guidelines for the preparation of radiopharmaceuticals in hospitals, London 1975 (Special Report No. 11).
5. Hospital Physicist Association. The Hospital Preparation of Radiopharmaceuticals in Hospitals (1976). Scientific Report Series-16.
6. Guidance notes for Hospitals. Premises and Environments for the preparation of Radiopharmaceuticals. Dept. of Health and Social Security, London 1982.
7. *Frier, M., S. R. Hesslewood:* Quality assurance of radiopharmaceuticals: a guide to hospital practice. Nuclear Medicine Communications, special issue, 1980.
8. Guidelines for radiopharmaceutical quality control in hospitals. (Rep 82EHD-77). Dept. Health and Welfare, Canada 1982.
9. *Robbins, P. J.:* Chemistry of Technetium-99m. Radiopharmaceuticals, A practical guide. Soc. Nucl. Med., New York 1984.
10. *Kristensen, K.:* First Report on the system for Reporting of Adverse Reactions and Drug Defects 1980–82 from the "Joint Committee on Radiopharmaceuticals of the European Nuclear Medicine Society and the Society of Nuclear Medicine, Europe". Nuklearmedizin **21** (1982), 274.
11. European system for reporting of adverse reactions and drug defects: Second report 1982–1983. Eur. J. Nucl. Med. **9** (1984), 388.
12. *Keeling, D. H., C. B. Sampson:* Adverse reactions to radiopharmaceuticals. United Kingdom 1977–1983. Brit. J. Radiol. **57** (1984), 1091.

Chapter 17

Quality Control and Drug Effect. Technical Aspects of Electrophoresis, Paper and Thin-layer Chromatography, Column Chromatography

Ingrid Kaempfer

1. Introduction

Medical scientists from all over the world have increasingly emphasized the importance of quality control for radiopharmaceuticals /1—12/.
Nuclear medical institutions use a great variety of radioactively labelled pharmaceuticals (Table 17.1.).
The spectrum used ranges from commercial radiopharmaceuticals over generator nuclides and technetium- and indium-labelled compounds (which are called Instant Kits) to radiopharmaceuticals synthesized at the clinic. The question arises: Which role should quality examinations play in the daily clinical routine of a nuclear medical institution? This question is difficult to answer. As far as commercial radiopharmaceuticals are concerned, the enclosed quality certificate of the manufacturer, which must be in line with the regulations laid down in the pharmacopoeia of the GDR, should suffice. A quality control seems to be necessary only then, if the medical doctor identifies deficiencies during application.
For this reason, a radiochemical laboratory at a nuclear medical clinic should be prepared to conduct, if necessary, analytical examinations for this first group of compounds. In this connection it is important to fall back upon standardized analysis methods in order to obtain comparable results at different institutions. A pharmacopoeia is being prepared.
We strive to use methods coordinated with the manufacturer. The manufacturer delivers compounds to several clinics to monitor the methods. The analysis results are then compared. All the other compounds, which are prepared in some way before application at a radiochemical laboratory, should in any case be subject to quality control.
This concerns generator nuclides, Tc- and In-labelled radiopharmaceuticals and, of course specific developments produced by a clinic. As the Tc- and In-labelled radiopharmaceuticals are manufactured only shortly before

Table 17.1. Radiopharmaceuticals

Group 1	commercial radiopharmaceuticals (i. e. I-131-, Cr-51-, Yb-169-, Tl-201-, Xe-133-compounds)
Group 2	Tc-99m- and In-113-eluants
Group 3	kit-prepared Tc-99m- and In-113-radiopharmaceuticals
Group 4	radiopharmaceuticals prepared at nuclear medical departments

application, practical, reliable, fast and, if possible, standardized analysis methods are required.
Nuclides, e.g. technetium-99m, exist in a carrier-free form, i.e. with a concentration of only around 10^{-9} up to 10^{-11} M. This means that these compounds can be only determined through gamma radiation.
As to the evaluation of quality, we differentiate between quality parameters, which are determined by physico-chemical methods, and those parameters, which can be determined only through biological tests (Table 17.2.).
Now we are going to consider the first group of parameters, which can be determined by physico-chemical methods.

Table 17.2. Quality control tests

I *Physico-chemicals tests*	II *Biological tests*
radiochemical purity	sterility
radionuclide purity	apyrogenicity
chemical purity	toxicity
particle size	
pH	

2. Radionuclide purity

It is defined by the share of the radionuclide's activity in the compound related to the overall activity. A radiopharmaceutical can, apart from the desired radionuclide, contain other radionuclides which depends upon the manufacturing technique. These can be both radioisotopes of the same element and of other elements. Such impurities lead to an increased radiation dose for patients. Besides impurities caused by the manufacture, also other factors can play a role. When using generator nuclides the co-elution of the long-lived parent nuclide, e.g. Mo-99 with the Tc-99m-eluate, can occur. We have to deal with this problem in daily routine.
How is radionuclide purity determined?
Each nuclide possesses a characteristic energy spectrum which can be recorded with a multi-channel spectrometer. Not every nuclear medical institution has such a spectrometer. For determining Mo in Tc-eluate, a very simple method has proved succesful in daily routine. As Tc has a low gamma radiation of only 140 keV, it can be shielded by a special lead insert. In this way only Mo can be determined with a higher energy of 740 keV (in the isotope calibrator). Control of a possible molybdene breakthrough should be performed daily. A share of 0.01 per cent is permissible. If an exceptionally high percentage is found, then an application will be forbidden. Time expenditure amounts to 2 minutes.

3. Chemical purity

It is defined by the percentage by weight of the compound compared to the weight of the total compound. It can always be due to impurities in the inactive original material. Thus the analytical determination is performed by the manufacturer. Nuclear medical establishments are interested only in the determination of aluminium in technetium-eluate, because the presence of soluble Al-compounds can cause difficulties in labelling erythrocytes, disphosphonate, DMSA, etc. even at concentrations above 2 ppm. In the Mo-Tc-generator aluminium-oxide is used as an absorber material for molybdene. During elution of technetium produced by decay with physiological sodium-chloride solution, aluminium (Al) can be possibly co-eluated. A semi-quantitative determination method is available within the range of concentration of 1 ppm.

4. Particle size

When using labelled particles from human serum albumin, size distribution is very important. A simple microscopical examination provides information on size distribution.

5. Radiochemical purity

It is defined by the activity share of the radionuclide, which exists in the chemical form declared. As radiochemical impurities are enriched in other organs, they disturb measurements of the organ to be examined through an increased background radiation. The activity to be applied necessarily should be increased.

This, in turn, results in an increase in the radiation dose of patients. This quality parameter supplies us with an important information before applying the radiopharmaceutical: it is of great importance for quality controls at nuclear medical establishments.

Chromatographic methods are available to us for determining radiochemical purity. This notion was coined by the Russian botanist Tswett in the last century. Chromatography is based upon the separation of the components of a mixture through the way of distribution between two phases due to solubility and adsorption. A clear differentiation of chromatographic procedures—pure distribution chromatography or pure adsorption chromatography—is not possible, because besides distribution also adsorption processes are involved to a smaller degree. We base the chromatographic procedures upon the aggregate state of the two phases when classifying them.

Due to technical facilities paper chromatography, thin-layer chromatography, gel-chromatography and electrophoresis are used at nuclear medical institutions (Table 17.3.).

Table 17.3. Chromatographic procedure

Adsorbents	Developing solvents	Procedure
solid cellulose	liquid	paper chromatography
solid silica gel plates	liquid	thin-layer chromatography
solid Bio-gel, sephadex	liquid	column chromatography (Gel-chromatography)
solid or liquid	gas	gas-chromatography
solid	liquid pressure	high pressure liquid chromatography (HPLC)

6. Paper chromatography

Chromatographic papers are manufactured from cotton-cellulose. The selection of the proper composition of the mobile phase depends upon the character of the substances to be separated. According to the direction in which the solvent flows we distinguish between rising and declining paper chromatography. The running speed of the running medium is lowered through variation of the papers, e.g. a tighter network of the cellulose-fibrils. During the chromatographic process the components are distributed between the stationary and the mobile phase depending upon the distribution coefficient (Figure 17.1.). Electrostatical forces of the stationary phase retain some components, while the mobile phase distributes them. This effect and the different solubilities of the single components cause the differing migrations.

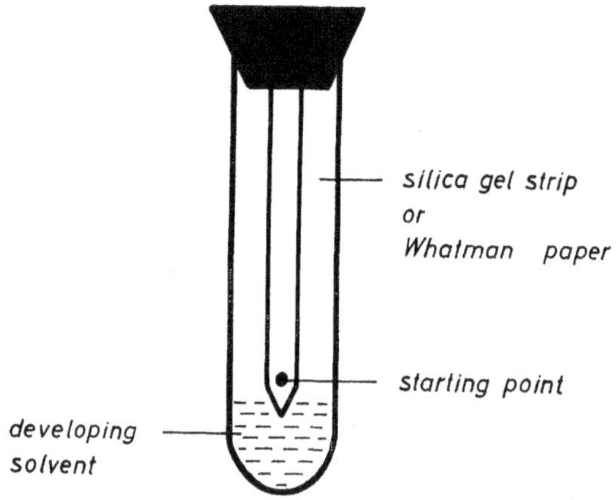

Fig. 17.1. Thin-layer- or paper-chromatography of Tc-compounds in organic solvent

The R_f-value (ratio of front) stands for the measure of this migration speed. It is a characteristic value for each compound in a certain solvent. The values can be only reproduced under defined working conditions:

- temperature constancy,
- purely mobile phase,
- equilibrating chromatographic papers at a constant period of time,
- well-closed vessel,
- specific paper quality.

7. Thin-layer chromatography

Thin-layer chromatography is based upon a similar mode of operation. Glas, plastic or thin metal plates being layered with carrier materials serve as the stationary phase. Numerous carrier materials with different physico-chemical characteristics are selected for layering the plates. Advantages over paper chromatography can be found in a lower time expenditure and partially in a preciser separation effect.

8. Gel-chromatography

It is based upon the separation of components according to the amount of their molecular weight.
The stationary phase is a gel, which consists of a largely inert macromolecular matrix, and which is saturated with a liquid. The solvent is part of the mobile phase.
Sephadexes, network polydextran gels or network polyacrylamide gels, the Bio-gels, are used. The separation can be schematically described as follows: Gels are differently permeable for different compounds. Permeability depends, on the one hand, upon the pore size of the gels used and, on the other, on the size and shape of the dissolved molecules. Molecules, that are bigger than the pores of the sephadex gels applied, do not permeate and thus always remain in the solvent phase. Consequently, they are first eluted. Smaller molecules permeate into the solvent spaces embraced by the gel depending on their size and shape. The gel more strongly retains smaller molecules than bigger ones. Elution, therefore, occurs depending on the order of decreasing molecular size.
There are gels of different pore size available for various possibilities of application.
Network polydextran-gels, i.e. sephadexes, even possess numerous free hydroxyl-groups, which can possibly interact with the compound to be investigated, which is an undesirable effect. Therefore, it is more favourable for gel-chromatography of weak complexes to use network polyacryl-amids, such as Bio-gel /13–16/. Persson developed the what is called gel-chromatography column scanning for examining Technetium-labelled compounds.

The columns are filled with the stationary phase. An aliquid part of the substance mixture is applied and after permeating into the gel interior it is eluted with so much eluate in order to obtain a maximum distribution of substances on the column. Then the evaluation can be performed. One obtains characteristic activity profiles (Figure 17.2.).

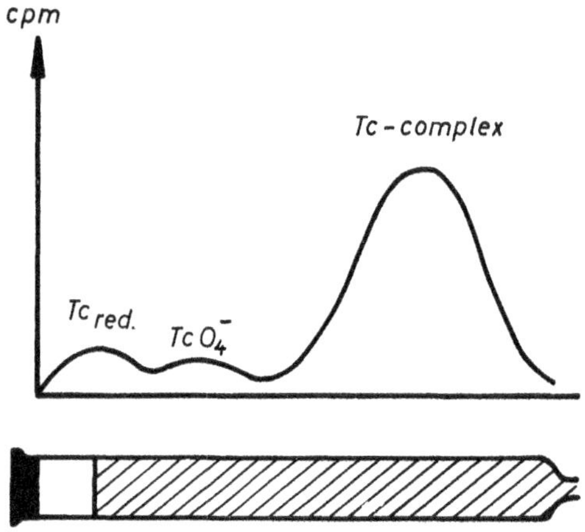

Fig. 17.2. Column-gel-chromatographic scanning of Tc-compounds

9. Paper electrophoresis

This procedure does not belong to the chromatographic methods. However, it has frequently been used for quality control at nuclear medical departments.
The principle of separation is based on different migration directions and migration speeds in the electric field.
According to definition, only those substances can be electrophoretically separated, which dissociate into the ion in the relevant solution. As a stationary phase we use a certain filter paper analogous to paper chromatography. The substance mixture is applied on a fixed spot of the paper, which was moistened with a buffer solution before. The paper strip is brought in such a way into the electrophoresis apparatus that the ends of the paper project into the buffer solution. With a voltage of 200 up to 300 volt, one obtains the desired separation after approximately one hour. This separation procedure, the what is called carrier electrophoresis, is no "real" chromatographic technique.
The electric field has the most important separation function, whose intensity, however, is essentially influenced by the characteristics of the stationary phase as well as the electrolyte solution.

10. Quality control performance

Evaluation of paper-, thin-layer-, column chromatography or electrophoresis is performed with the thin-layer scanner. The measured results supply information about the percental composition of a substance mixture and make it possible to determine the R_f-values. However, they do not allow any conclusions on the chemical structure of the compounds examined.
The choice of method depends upon the analytical approach.
The following schematic representation of a labelling process with the nuclide Tc-99m in Figure 17.3. can explain this.
This labelling is a very complex process. The following products can be obtained in this reaction:

1. free pertechnetate,
2. hydrolyzed, reduced technetium,
3. Tc-complexes of different composition,
4. polymeric products.

Since such impurities affect also the biological distribution of the radiopharmaceutical /17/, it is of great importance to work out an optimal labelling method in order to obtain a reproducable radiopharmaceutical. For this reason, instant kits are developed by the manufacturers. These instant kits contain the desired inactive ligand-reducing agent complex in a lyophiliced form. The proportion of the ligand to the reducing agent and Tc-generator type is experimentally obtained in such a way that a reproducable radiopharmaceutical develops through addition of Tc-generator eluate after a certain incubation period. Although the labelling process with the instant kit has been optimized by the manufacturers, there are many influential factors which require an examination of the finished radiopharmaceutical to be applied.

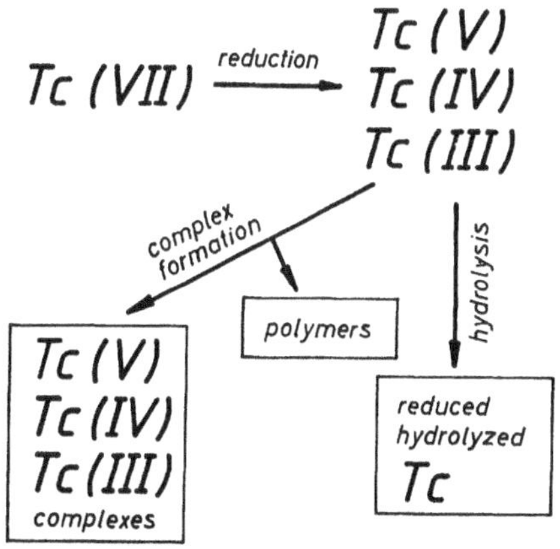

Fig. 17.3. Labelling scheme of Tc-compounds

1. Concentration of stannous ion
   Insufficient stannous ions are available for the complete reduction of technetium due to the oxidation of Sn in the vial of instant kit.
2. pH variety.
3. Complex stability in vitro and in vivo
   Technetium complexes meet in vivo with various chelating agents during their transport. A moderate technetium complex can easily convert into other complexes with higher stability.
4. Another problem of radiochemical purity is connected with the use of kits and technetium generators, if generator and kit come from different manufactures.
5. If a generator is not used for a long period of time, especially on Monday, the labelling with the first eluate is not optimal.

Occasionally, the quantities of free technetium-99m and some forms of hydrolyzed technetium-99m in our radiopharmaceuticals have exceeded permissible limits.
It is possible to determine free pertechnetate by thin-layer or paper chromatography. Free pertechnetate closely migrates to the acetone solvent front, whereas specific technetium radiopharmaceutical remains at origin, and so does hydrolyzed-reduced technetium, too.
If saline is used as a solvent, hydrolyzed-reduced technetium remains at origin, whereas free pertechnetate and specific technetium radiopharmaceutical migrate with the saline solvent front.
Separation is optimal and thin-layer chromatography procedures are extremely rapid—about 10 up to 15 min.
We use this chromatography system in the daily routine of nuclear medical departments.
A determination of hydrolyzed-reduced technetium, free pertechnetate and specific technetium radiopharmaceutical is also possible by gel chromatography column scanning. If no interaction takes place, reduced hydrolyzed technetium remains on the top of the column, then free pertechnetate migrates followed by the specific radiopharmaceutical.
For some radiopharmaceuticals such as HIDA electrophoresis is the method of choice. This procedure is slow; it takes about 60 minutes. In the daily routine at our nuclear medical department we investigate the following radiopharmaceuticals: Tc-99m-EHIDA, Tc-99m-microspheres, Tc-99m-diphosphonate, Tc-99m-DMSA, Tc-99m-HSA.
The structure of the compounds cannot be estimated by these analytical methods. A distinction between various technetium complexes is only possible in the biological system.
Distribution studies in animals are necessary for applying new radiopharmaceuticals, e.g. in research, but not for clinical routine.
After a testing period of 3 up to 4 years, we can say that about 1 per cent of all compounds applied shows quality defects. Investigations into parameters such as sterility and apyrogenicity are very time-consuming. They cannot be performed under the conditions of daily routine. In this case the quality certificate of the manufacturer has to be used.

However, if a sterilization is required, then sterile filtration with millipore filters should be performed. If, however, at the clinic radiopharmaceuticals are produced with an exemption permit issued by the central drug institute (Arzneimittelinstitut), then the biological tests have to be carried out by all means.
In summary, we can say that the labelling process as well as the percentage of the constant of labelling ingredients can be determined with simple, rapid and reliable methods.
This quality control should become routine for all radiopharmaceuticals manufactured at a clinic, whereas commercial products should be analyzed only then, if the doctor recognizes defects during the investigation. The development and testing of new radiopharmaceuticals requires many more detailed investigations. As we do not determine the structure of the compound with the analysis procedures described above, also the identification of the compound cannot be detected. Thus, also the biological behaviour cannot be definitely predicted. In such cases, differentiation is only in the biological system, i.e. in animal experiments, possible.

References

1. *Heide, L., A. Stamm, W. Bögl:* Analytik radiochemischer Verunreinigungen in Radiopharmaka. Dietrich Reimer Verlag, 1980.
2. *Frier, M., S. R. Hesslewood:* Quality Assurance of radiopharmaceuticals. Chapman and Hall, 1980.
3. *Kristensen, K.:* Quality control analysis at the hospital. Radiopharmaceuticals II, Proc. 2. Int. Symposium, Seattle 1979.
4. *Eckelman, W. C.:* Radiochemical purity of new radiopharmaceuticals. J. Nucl. Med. **17** (1976), 865.
5. *Pauwels, E. K. J., R. I. J. Reitsma:* Radiochemical quality control of Tc-99m-labeled radiopharmaceuticals. Eur. J. Nucl. Med. **18** (1977), 97.
6. *Marzilli, L. G., P. Worley, H. D. Burns:* A new electrophoretic method for determining ligand. J. Nucl. Med. **20** (1979), 871.
7. *Zimmer, A. M., D. G. Pavel:* Rapid miniaturized chromatographic quality-control procedures for Tc-99m radiopharmaceuticals. J. Nucl. Med. **18** (1977), 1230.
8. *Zimmer, A. M., R. A. Holmes:* A precise chromatography system for specific Technetium-99m radiopharmaceuticals. Nucl. Med. **14** (1975), 805.
9. *Billinghurst, M. W.:* Chromatographic quality control of Tc-99m-labeled compounds. J. Nucl. Med. **14** (1973), 793.
10. *Gutkowski, R. F., H. J. Dworkin:* A simplified radiochromatographic purity check. J. Nucl. Med. **12** (1971), 513.
11. *Colombetti, L. G., S. Moelien, G. C. Patel* et al.: Rapid determination of oxidation state of unbound Tc-99m and labeling yield in Tc-99m labeled radiopharmaceuticals. J. Nucl. Med. **17** (1976), 805.
12. *Albers, J. W., D. Jenkins, R. J. Sandeen* et al.: Free unreacted pertechnetate in technetium-sulphur colloid preparations. J. Nucl. Med. Techn. **2** (1974), 14.
13. *Persson, B. R. R., L. Darte:* Gel chromatography column scanning for the analysis of Tc-99m-labeled compounds. J. Chromatogr. **101** (1974), 315.
14. *Johannsen, B.:* Gelchromatographische Qualitätskontrolle von Tc-99m-Radiopharmaka. Radiobiol. Radiother. **17** (1976), 91.

15. *Johannsen, B.:* Behaviour of Tc-99m-labeled diphosphate on Sephadex and Bio Gel. J. Nucl. Med. **16** (1977), 10—87.
16. *Steigman, J., H. P. Williams et al.:* Gel chromatography in the analysis of Tc-99m-radiopharmaceuticals. J. Nucl. Med. **15** (1974), 318.
17. *Hambright, P., J. McRae* et al.: Chemistry of technetium radiopharmaceuticals. J. Nucl. Med. **16** (1975), 478.

Chapter 18

Adverse Reactions to Radiopharmaceuticals

Harald Deckart

Shani et al. described the adverse reaction as an "unanticipated physiologic response of the patient to the vehicle carrying the radiation, not the radiation itself". Radiation effects should be properly categorized as an overdose problem; we have observed this for instance as necrosis and skin ulcerations on the injection site after paravenous application of ^{201}Tl-chloride /10/, the subcutaneous injection of ^{198}Au-colloid for lymphscintigraphy, para-articular injection of ^{90}Y-colloid in radiosynovectomy; gastrointestinal symptoms after oral application of ^{131}I for thyroid treatment are known as radiation effect just as fibrosis and inflammation after intraperitoneal application of ^{198}Au-colloid for treatment.
Adverse reactions of radiopharmaceuticals are due to the pharmacologic effects of carriers or stabilizers. Symptoms are either *allergic* or idiosyncratic (hives, bronchospasm dyspnoea, shocklike state, histamine-type reaction such as flushing, subjective itching of hands and feet, hypotension) or *toxic pharmacologic* effects (vasoconstriction, headache, backache, dyspnoea) or due to *pyrogens* or *nonsterility* (infection, fever) or local effects on the injection site (thrombophlebitis, vascular occlusion by clots of particles, as can be seen in Table 18.1.

1. Reporting systems

Since 1967 the Society of Nuclear Medicine (SNM-USA) has recorded cases with adverse reactions and published various summaries /1, 6, 9, 10/. In 1976 a "SNM Drug Problem Reporting System" was established as a cooperative arrangement between the Society of Nuclear Medicine, the US Pharmaco-

Table 18.1. Type of adverse reaction

	Pyrogens Nonsterility	Vasovagal toxic or pharmacological	Allergic or iodosyncratic	Meningism
USA-Reports	5	51	25	68
GDR	5	5	51	10
UK	not clearly to be differentiated from literature			
total: 220	10	56	76	78

poeia Convention (USP) and the Food and Drug Administration (FDA) /6/, which replaced the former SNM adverse reaction registry. In the UK a reporting system was set up by the Nuclear Medicine Committee of the British Institute of Radiology, the British Nuclear Medicine Society and the Hospital Physicians Association in 1972 /7/. In the GDR a reporting system was set up in 1973 /5/, by the Society of Nuclear Medicine (the former Section of Nuclear Medicine in the Society of Radiology) jointly with the Institute of Drugs. Every nuclear medical department was obliged to send in the case of any adverse reaction a questionnaire to the Society's secretary and a sample of the injected material for quality control, pyrogen testing and bacteriological examinations to the Institute of Drugs.
Since 1980 the European Joint Committee on Radiopharmaceuticals has sought to collect reports on adverse reactions and drug defects from all European countries; however, only few countries have been participating in this project till now /2/. This may be due to the lack of centralized reporting systems in most of the countries.

Form of reporting

Under the above-mentioned reporting systems questionnaires have been developed. The reporting system of the GDR contains information about: address and phone number of the nuclear medical centre concerned, characterization of the radiopharmaceutical used: name, lot-number, code, manufacturer, date of preparation, day and kind of application, applied volume and activity, ingredients and pH of the (home made) preparation, results of quality control carried out at the radiopharmacy of the clinic, intended kind of investigation, patient's sex and age, detailed diagnosis, anamnestic data on allergic hypersensibility, characterization and kind of adverse reaction, time of onset and duration after application, way of application, kind and efficiency of therapeutic intervention, the examining physican's opinion on a possible connection between reaction and patient's disease(s), the examining physician's signature.

2. Experience from reporting system

Table 18.2. summarized published adverse reactions from the USA (1967 to 1981), from the UK (1973–1983) and from the GDR (1974–1983). A total of 621 reports is summarized from the USA (490), the UK (61), the GDR (70). The European Report of adverse reactions contains from 1980 to 1983 altogether 75 cases; 58 of them are included in Table 18.2. (UK: 33; GDR: 25); except the rest of 17 reports collected from Switzerland (1), Denmark (7), Italy (1); The Netherlands (7); Finland (1). In the GDR 65 % were of minor or medium intensity, 25 % were severe reactions requiring intensive care, careful treatment and observation at a hospital; 1 case was lethal (without clear cause-relationship: reactive pulmonary emboli at the moment of intravenous application of ^{51}Cr-labelled RBCs).

Table 18.2. Frequency of adverse reactions to different radiopharmaceuticals in three countries

Radiopharmaceutical	Total	USA	UK	GDR
^{131}I-HSA	73	72	1	
Cholesterol	33	7		26
Hippurate	17	11	3	3
NaI	16	16		
Rose bengale	4	4		
^{123}I-NaI	7	7		
IMP	1		1	
^{99m}Tc-Sn-colloid	119	99	18	2
microspheres	79	71	4	4
MAA	20	16	4	
ferrous hydroxide MAA	32	32		
MDP	31	13	17	1
pyrophosphate	16	16		
DTPA	22	21	1	
pertechnetate	16	8	1	7
HSA	7	7		
polyphosphate	5	5		
DMSA	4	3	1	
glucoheptonate	10	7	1	2
^{113m}In-ferrous hydroxide	42	35		7
^{111}In-DTPA	14	13	1	
^{198}Au-colloid	11	4	2	5
^{169}Yb-DTPA	5			5
^{203}Hg (^{197}Hg)-Chlormerodrine	4	3		1
^{32}P-orthophosphate	4	4		
^{85}Sr-chloride	3	3		
^{67}Ga-citrate	3	1	2	
^{18}F-Al-complex	3			3
^{75}Se-cholesterol	3		1	2
^{51}Cr-RBC	2		1	1
^{111}In-tropolone platelets	1		1	
^{201}Tl-chloride	1	1		
^{87m}Sr-chloride	1	1		
^{90}Y-colloid	1			1
^{59}Fe-desferrioxamine	1		1	
total	621	490	61	70

Characterization of the type of adverse reaction connected with special radiopharmaceuticals

Labelled colloids (20.9 % adverse reactions of the sum total)
Reports on side-effects on the primarily used ^{198}Au-colloid for RES-scintigraphy were very rare (1.8 % in this series), but they increased to 19.2 % with ^{99m}Tc-S-colloid. Most of the reactions reported were caused allergotoxically by stabilizing agents: HSA, gelatine, dextran, carboxymethylcellulose, mannitol, in particular in higher concentrations. Deaths have occurred

especially in the state of severe disease. Successful trials without side-effects have been made in freshly-used preparations without applying stabilizers. Though the incidence of adverse reactions has decreased in the last 5 years, colloids have been reported with a proportion of 33 % in the USA /2/, 25 % in the UK /7/ of all registered reactions, while the GDR has reported only 3.7 % /5/.

Labelled phosphates/phosphonates (8.7 % adverse reactions of the sum total)
In the last decade an increasing proportion of adverse reactions to bone-seeking radiopharmaceuticals has been recorded, in particular to MDP (5.0 % in this series); in contrast to this, only 1.1 % side-effects have been published as far as the formerly used radiopharmaceuticals are concerned: ^{85}Sr, ^{87m}Sr, ^{18}F. After MDP-injection a skin rash has been observed which occurs maculopapularly, erythematously and puritically; it develops during 1 to 48 hours and it disappears after applying antihistaminic drugs or after corticosteroid treatment /7, 11/.

Labelled cholesterols (5.8 % adverse reactions of the sum total)
Immediately after the injection of ^{131}I-labelled cholesterol derivatives (19-Iodocholesterol, Adosterol®), we observed in 1974 allergotoxic reactions in medium and partly severe intensity: 26 out of 33 reported cases have been observed in the Nuclear Medicine Clinic at Berlin-Buch since 1975, although we preinjected 10 mg Prednisolone intravenously just before injecting the radiopharmaceutical very slowly. Vein crampi, headache, flushes, hot feeling of the skin, tachycardia, hypotension were symptoms, less observed in ^{75}Se-labelled cholesterol (Scintadren®). Side-effects were also reported from Japan with a rate of unknown incidence /8/.

Particulated albumin (15.9 % of all reports)
In the 1960's macroaggregated albumin (MAA) was mixed in the syringe with blood before injection; aggregation with blood clots occurred leading to a mechanical obstruction of capillaries which resulted in localized embolism with protracted biodegradation. Several fatalities were described /10/ in particular in patients with advanced pulmonary vascular disease. We observed long-term deposit in the iugular vein /5/. Apart from this, anaphylactic reactions due to ferric chloride in the solution were observed.
There was an increase after changing from ^{131}I- or ^{99m}Tc-labelled MAA to -microspheres, perhaps relatively, as these radiopharmaceuticals were used more frequently in cardiac and circulatory studies.

Iron hydroxide precipitates (11.9 % of the sum total)
In the 1960's at first ^{113m}In, then ^{99m}Tc-labelled ferrous hydroxide macroaggregates were used for lung perfusion scintigraphy stabilized with gelatine. Microgram quantities of free iron (Fe^{3+}) induced toxic reaction with vasoconstriction; adverse reactions were also due to gelatine and to conglomeration of particles. Immediately after application severe allergotoxic reactions were observed, many deaths have been reported. Therefore, this radiopharmaceutical was withdrawn in the mid-1970's.

Human serum albumin (12.9 % adverse reactions of the sum total)
Most adverse reactions were reported after intrathecal administration: asep-

tic meningitis occurred with headache, meningism, fever due to high albumin content and/or pyrogenes. Endotoxin is 1,000 times more toxic in the cavum leptomeningicum comparable to the blood pool. The rate of incidence decreased after the introduction of the limulus test, which is much more sensitive in detecting pyrogens compared to the rabbit test. Labelled HSA has been now replaced by ^{111}In-, ^{99m}Tc- or ^{169}Yb-labelled chelates (DTPA) for cisternography.

^{131}I-hippuran (2.7 % adverse reactions of the sum total)
After the application of hippuran local thrombophlebitis has been observed in 2 out of 6 cases (UK, GDR); others have shown vasovasal reaction with hypotension.

Other radiopharmaceuticals
After intravenous injection of the solutions of Na^{131}I, Na^{123}I (3.7 %), ^{99m}Tc-pertechnetate (2.6 %), ^{99m}Tc-DTPA (3.5 %), ^{99m}Tc-glucoheptonate (1.6 % of the sum total) different, mostly allergic reactions of low incidence have been reported; some contained pyrogens or were unsterile. The latter can be nearly totally avoided, since modern preparation techniques (laminar flow box, etc.) and quality control (sterility tests, limulus test) were performed in the industrial production of radiopharmaceuticals and labelling kits (see Chapters 15, 16).
Special attention has to be given to preparations for intrathecal application: 10 of our patients (GDR) developed meningitis after intrathecal injection of ^{169}Yb-DTPA for cisternography and ^{198}Au-colloid for prophylaxis against meningosis leucaemica /4/. These radiopharmaceuticals might not have had any certificate for intrathecal use (limulus test, see above). ^{90}Yb-colloid and ^{186}Re-colloid preparations induced inflammation after introduction into-knee- or wrist-joints for treatment (radiosynovectomy): the cause could not be clarified exactly.

3. Incidence

As far as the early and late periods of the reports on adverse reactions from these three countries are concerned, the incidence was reduced in the second period which is due to an increased knowledge in the cause-relationship and due to standardized and controlled production techniques and the introduced quality control (Table 18.3.). The high incidence of side-effects due to labelled cholesterol is considerable (Table 18.4.).
The incidence of adverse reactions to radiopharmaceuticals varying from 0.001 to 0.037 % (Table 18.3.) has to be compared to the incidence of X-ray contrast media: intravenous cholangiography (7.5 to 15.3 %), intravenous urography (4.6 to 8.5 %), angiography (0.4 to 3.6 %) /7; 12/.

4. Conclusion

- The probability of an adverse reaction to the radiopharmaceutical is about 1/100 of that of a therapeutic drug or contrast medium.

Table 18.3. Incidence of adverse reactions to radiopharmaceuticals in various countries

Country	Period of report	Reactions per 100,000 examinations	
USA	1967—1970	20—100	estimated
	1976—1981	1— 6	
Japan	1977—1979	6— 20	
UK	1973—1983	37	(estimated)
GDR	1974—1979	2.9	
	1980—1983	1.6	
	1974—1983	2.2	

Table 18.4. Incidence of adverse reactions to labelled cholesterol derivatives

Country	Radiopharmaceutical	Incidence		
Japan	labelled cholesterol	3—5 reactions / 1,000 examinations		
GDR	labelled cholesterol	59.9	—	—
	^{131}I-cholesterol	69.0	—	—
	^{75}Se-cholesterol	23.3	—	—

- The incidence all over the world has been significantly reduced (roughly by the factor of 20) during the last decade with respect to better controlled preparation techniques and generally used quality control.
- Many of the earlier reactions have been attributed to ironcontaining preparations, gelatine-stabilized products or materials, such as HSA, contaminated with pyrogens. Many of these materials involved in earlier reactions are now no longer used in radiopharmaceutical formulations.
- Considerable improvements in radiopharmaceutical manufacturing and quality control during the past decade have reduced the incidence of adverse reactions; no deaths were attributed (compare chapters 15, 16 and 17) to the application of radiopharmaceuticals.

References

1. *Atkins, H. L., W. Hauser, P. Richards, J. Klopper:* Adverse reactions to radiopharmaceuticals. J. Nucl. Med. **13** (1972), 231.
2. *Cordova, M. A., W. B. Hladik, B. A. Rhodes:* Validation and characterization of adverse reactions to radiopharmaceuticals. Noninvas. Med. Imag. **1** (1984), 17.
3. *Cox, P. H.:* Personal communication.
4. *Deckart, H.:* Nuklearmedizin in Theorie und Praxis. 2. Edition. Gustav Fischer, Jena 1979, p. 364.
5. *Deckart, H.:* Adverse reactions to radiopharmaceuticals. Reporting System in the GDR 1973—1980. Europ. Nucl. Med. Congress, Pisa 1981.

6. *Ford, L., A. Shroff, W. Benson, H. Atkins, B. A. Rhodes:* SNM Drug problem reporting system. J. Nucl. Med. **19** (1978), 116.
7. *Keeling, D. H.:* Adverse reactions to radiopharmaceuticals. United Kingdom 1977–1983. Brit. J. Radiol. **57** (1984), 1091.
8. *Kristensen, K.:* Reporting of adverse reactions and drug defects. Europ. J. Nucl. Med. **9** (1984), 45.
9. *Rhodes, B. A., M. A. Cordova:* Adverse Reactions to Radiopharmaceuticals: Incidence in 1978 and associated Symptoms. Report of the Adverse Reactions Subcommittee of the Society of Nuclear Medicine. J. Nucl. Med. **21** (1980), 1107.
10. *Shani, J., H. L. Atkins, W. Wolf:* Adverse reactions to radiopharmaceuticals. Semin. Nucl. Med. **6** (1976), 305.
11. *Spicer, J. A., D. F. Preston, R. L. Stephens:* Adverse allergic reaction to technetium-99m methylene diphosphonate. J. Nucl. Med. **26** (1985), 373.
12. *Shehadi, W. H., G. Toniolo:* Adverse reactions to contrast media. Radiology 137 (1980), 299.

Chapter 19

Pharmaceutical Interaction with Radiopharmaceuticals: I. Radioiodine-, Pertechnetate Kinetics in Thyroid Endocrinology

Harald Deckart

In the last decade several interference of drugs with iodine kinetics and thyroid hormone metabolism have been reported /1, 7, 11, 25/. As far as the complicated regulation system is concerned, those interferences may influence only with one position or may interfere with different points in the long chain of the thyroid system in the axis of central nervous system → hypothalamus → pituitary → thyroid gland → hormone transport system on the blood stream → peripheral thyroid hormone degradation. A great variety of drugs used as analgetics antirheumatics, antiinflammatory drugs, tranquillizers, narcotics, antiepileptic drugs, hormones and cardiac glycoside has been recognized to interfere with hormone metabolism. As a result, changes were obtained in radioiodine kinetics and in the concentration of radioimmunologically examined hormone concentrations which are not due to a pathomechanism in the thyroid hormone system. These drug and drug-dose dependent changes have to be strictly considered by the examining physician, the endocrinologist or the general practitioner.

1. Hypothalamus – pituitary axis

1.1. Stimulation with an *increase of TRH/TSH-secretion* was observed after:

Dopamine blocking agents (metoclopramide, sulpiridine, haloperidol, biperidine, chlorpromazine), the sexual hormone progesteron, the aldosteron-antagonist spironolactone, iodine, iodine-containing contrast media and the antiarrhythmic drug amiodarone, the otnopic psychotropic drug lithium, the vasodilatator theophylline /15, 19, 22, 24/.

1.2. Depression with a *decrease of TRH/TSH-secretion:*
TRH-antagonist dopamine, l-DOPA (drug against Parkinson syndrome), the coenzyme of dopamine-synthesis pyridoxine (vitamin B), the dopamine-antagonist bromocryptine, used in hyperprolactinemia, sterility treatment and acromegaly, the analgetic and antirheumatic drug phenylbutazone /13, 23/, the emetic and dopaminergic agonist apomorphine and opiates /5/, glucocorticosteroids /11, 12/ and somatostatin /25/.

2. Thyroid

Drugs influence the complicated iodine kinetics at different levels.

2.1. *Iodine-transport:* the accumulation of I^- through the cell membrane into the follicular cell (characterized by radioiodine uptake test (RUT) can be inhibited by

- high I^- plasma concentration (borderline 20 μg/cl),
- the monovalent anions of the halogens Br^-, ClO_4^-, as the non-halogen TcO_4^- with their spherical shape,
- the linear-shaped pseudohalogenic anions SCN^-, $SeCN^-$.

2.2. *Iodination,* the second step including I^--oxidation catalized by thyroid-peroxidase (TPO) and the organification of iodine amino acids into mono- (MIT) and diiodotyrosine (DIT) can be interfered in different ways:

- temporary inhibition of TPO-reaction by iodide of high concentration (30 μg/cl), the Wolff-Chaikoff-effect /26, 27/, leads to a decrease in ^{131}I-uptake in RUT /6, 7, 19/. This effect is a temporary adaptation process to prevent an increase of thyroid hormone synthesis.

In euthyroids a unique application of iodine-containing X-ray contrast media (3 to 28 g organically bound iodine) beginning after 6 weeks and covering a period of months, an elevation of l-T_3-plasma level, and a decrease in basal TSH-level was observed as a result of a persisting Wolff-Chaikoff-effect with a low TRH-response after passive diffusion of iodine into the thyroid cell and a change in the normal MIT/DIT-ratio. These high doses of iodine-contamination, often occurring in clinical practice nowadays, lead to disturbances of the autoregulatory system, as described by Ingbar /14/.

- TPO can be temporarily inhibited by thionamides (thyrostatic drugs: propylthiourazile (PTU), methylthiourazile (MTU), carbimazole, methimazole), thiocyanate (SCN^-), p-aminobencoic acid, ascorbinic acid, Vitamin C), cysteine, sulphonamides, sulphathiazole), phenylbutazone.

2.3. *Condensation* of MIT and DIT into triiodothyronine (T_3) and thyroxine (T_4) is inhibited by thionamides. Thyroxin peroxidase (TPO)-inhibiting drugs are effective in the condensation process just as in the iodination step (2.2.).

2.4. Thyroid hormone secretion from the follicular cell is inhibited by lithium, chlorpromazine, propranolol, and by iodine-excess of extremely high doses ($>$ 30 μg/cl) /9, 15, 18, 21/.

3. Thyroid hormone transport

Different drugs influence the concentration of thyroxine-binding proteins (TBP), especially thyroxine-binding globulin (TBG) by competitive binding: the antiepileptic drugs carbamazepine, phenytoin, primidone /4, 10/, the analgetic and antiphlogistic drug phenylbutazone and fenlofenac /3, 13, 23/,

the antilipemic drug halofenate, the tranquillizers diazepam, chlordiazepoxide, the anticoagulans heparine, the antiphlogistic drug salicylate and sulphonyluric acid /8, 11, 17, 24/.
An *increase in TBG-production* is induced by estrogens, in particular in women who regularly take the antibaby pill /24/, while the antilipemic drug Clofibrate and opiates increase TBG, androgens and fluorazile (cytostatic drug) just like glucocorticosteroids lead to a *decrease in TBG-synthetization*. The drug-induced TBG-changes may lead to increased/decreased T_3- and T_4-concentrations and false-positive or false negative interpretation except for simultaneously estimated TBG-concentration with RIA-technique.

4. Peripheral hormone metabolism

Two thirds of l-T_3 are derived from metabolic degradation $T_4 \rightarrow T_3$ in peripheral tissue, called the T_4/T_3-conversion. It was observed that this process was influenced by different drugs. Phenytoine, carbamazepine, phenobarbital increase *T_4/T_3-conversion*, may be diagnosed falsely. The β-blocking propranolol, glucocorticosteroids and X-ray contrast media have been observed to decrease the *T_4/T_3-conversion rate* /1, 6, 18, 19/.
The widespread use of chronic amiodarone administration leads in euthyroid subjects to an increase in T_4-, free T_4- and reverse T_3-concentration in plasma and to a decrease in l-T_3 plasma level: it *blocks the metabolic action of thyroid hormones in peripheral tissue* and *inhibits peripheral conversion* of T_4 to T_3. Amiodarone contains 37.2 % iodine per weight; during longterm antiarrhythmic treatment and prevention of resistant supraventricular and ventricular arrhythmias 600 to 1,200 mg were applied per day /19/. This drug has an extremely slow degradation with a half-life of about two weeks; i.e. also after stopping treatment one has to take into account an interference covering 2 up to 4 weeks.

5. Iodine Excess

Firstly, after iodine treatment or as unintended side-effect in diagnostic procedures, intrathyroidal I^--concentrations are observed with an enhancement of 10^3 compared to the physiological level (13 mg/cl). In acute situation the I^--transport, p.e. the I^--uptake (RUT) may be not influenced by the I^--organification (2.3.).
In low cellular I^--content or after a perchlorate induced I^--efflux nothing will happen. In high I^--cellular concentration one may observe the transient inhibition of the I^--organification (Wolff-Chaikoff-effect, see 2.2.). It is succeeded by an adaptation process the so called "escape phenomenon": a decrease in I^--transport and the disappearance of the Wolff-Chaikoff-effect. Secondly, extremely high doses cause a qualitative change in the synthesis

of iodine amino acids: instead of the synthesis of metabolic active T_3 and T_4 an increase of the precursors MIT and DIT occurs, especially an increase in the MIT/DIT-ratio and a decrease in I^--reutilization after MIT-, DIT-deiodination. Thirdly, extremely high doses lead to an inhibition of Tg-proteolysis, an indirect inhibition of hormone secretion.

6. Influence of I-131-thyroid uptake

- opiates, heroin addicts: the iodine uptake test with ^{131}I has been generally withdrawn in the last decade from radiation hygiene aspects. It was replaced partially by ^{123}I-uptake, especially by ^{99m}Tc-trapping test, which can be parallelly recorded during thyroid scanning. Nowadays drug interference with iodine-kinetics plays an important role in the detection and treatment of thyroid cancer and its metastases.
- decrease by administration of *glucocorticosteroids* with marked differences depending on the route of application (oral, intravenously) /11, 12/.
- decrease by administration of *phenylbutazone* /13, 23/ as early as the first week of treatment in euthyroid and hyperthyroid patients.
- slight increase in iodine uptake by *fluoride*-containing drugs, especially toothpastes and a significant increase in chronic industrial fluorosis /2/.

7. Multiple effect of drugs

Some examples may demonstrate the *multiple effects of drugs* on the pituitary-thyroid-peripheral axis:

- *Glucocorticosteroids* decrease TSH-secretion and TRH-effect, decrease peripheral T_4/T_3-conversion with an increase in reverse T_3-level, decreased l-T_3-plasma concentrations and decrease in TBG-production, and a reduced iodine uptake in the thyroid (RUT).
- *Phenylbutazone* decreases TRH/TSH-secretion and inhibits iodine uptake of thyroid; secondly, it displaces T_4 from TBG-binding sites followed by an increased free T_4-level. In chronic treatment a normalization has been observed. The administration of this drug shows an inhibition effect on TPO, which influences the MIT/DIT-condensation mechanism similar to the action of thyrostatic drugs (thionamides).

References

1. *Breuel, H.-P., Ch. Breuel, D. Emrich, P. Fischer, A. Gerlak, C. Winkler:* Veränderungen der Schilddrüsenfunktion durch Röntgenkontrastmittel bei Schilddrüsengesunden. Med. Klin. **74** (1979), 1492.
2. *Bürgi, H., L. Siebenhüner, E. Miloni:* Fluorine and thyroid gland function: A review of the literature. Klin. Wochenschr. **62** (1984), 564.
3. *Capper, S. J., M. J. Humphrey, A. B. Kurtz:* Inhibition of thyroxine binding

to serum proteins by fenlofenac and related compounds. Clin. Chem. Acta **112** (1981), 77.
4. *Cavalieri, R. R., A. G. Laurence, A. Wallace, M. E. Hammond, K. Cruse:* Serum thyroxine free T_4, triiodothyronine and reverse-T_3 in diphenylhydantoin treated patients. Metabolism **28** (1979), 1161.
5. *Chan, V., Ch. Wang, R. T. T. Yeung:* Effects of heroin addiction on thyrotropin, thyroid hormones and prolactin secretion in men. Clin. Endocrinol. **10** (1979), 557.
6. *Ceccarelli, C., E. Grasso, E. Martino, M. Nardi:* Reduction of thyroid uptake by iodine absorbed with eye-drop therapy. J. Nucl. Med. **23** (1982), 364.
7. *Deckart, H.:* Die Schilddrüse. Aktuelle funktionsmorphologische Diagnostik. Schriftenreihe: Anwendung von Isotopen und Kernstrahlungen in Wissenschaft und Technik, Nr. 14, ISSN 0233—1268, Berlin 1984.
8. *El-Hazmi, M. A. F.:* Aspects of the interaction of thyroxine and its binding serum proteins with 1,4-Benzodiazepine tranquillizers. Biochem. Med. **26** (1981), 191.
9. *Faber, J., T. Friis, C. Kirkegaard, L. B. Lumholtz, J. Mølholm Hansen, K. Siersaek-Nielsen, L. Skovsted, P. Theilade:* Serum T_4, T_3 and reverse T_3 during treatment with Propranolol in hyperthyroidism, L-T_4 treated myxedema and in normal man. Horm. Metab. Res. **11** (1979), 34.
10. *Fichsel, H.:* Nebenwirkungen der Antiepileptika bei Langzeitbehandlung. Diagnostik **10** (1977), 599; Klin. Pädiat. **188** (1976), 435.
11. *Flierdt, E. Vande, E. Reich, W. Böge:* Interaktionen zwischen Radiopharmaka und Therapeutika: Tierexperimentelle Untersuchungen zur Beeinflussung der Pharmakokinetik von Technetium-Verbindungen durch Acetylsalicylsäure, Diazepam und Prednisolon. ISH-Heft 37, ISSN-0175-4254 (München 1984).
12. *Gamstedt, A., G. Järnerot, B. Kågedal, B. Söderholm:* Corticosteroids and thyroid function. Acta Med. Scand. **205** (1979), 379.
13. *Hansen, J. E. M.:* Mechanism of the effect of phenylbutazone on the thyroid gland. Acta Endocrinol. (Kbh.) Suppl. 67 (1962), 162.
14. *Ingbar, S. H.:* Autoregulation of the thyroid. Response to iodide excess and depletion. Mayo Clin. Proc. **47** (1972), 814.
15. *Kleinau, E., H. Deckart:* Optimierung der funktionellen Radiojodresektion von Hyperthyreosen durch Lithiumprämedikation. In: Radioaktive Isotope in Klinik und Forschung. Höfer, R. (ed.). Vol. 14/2, Egermann, Wien 1980, p. 477.
16. *Kristensen, B. O., J. Weeke:* Propranolol-induced increments in total and free serum thyroxine in patients with essential hypertension. Clin. Pharmac. Therapeut. **22** (1977), 864.
17. *Larsen, P. R.:* Salicylate-induced increases in free triiodo-thyronine in human serum evidence of inhibition of triiodo-thyronine binding to thyroxine-binding globuline and thyroxine-binding prealbumin. J. Clin. Invest. **51** (1972), 1125.
18. *Lumholtz, I. B., M. Busch-Sørensen, J. Faber, Th. Friis, G. Kirkegaard, K. Sierbaek-Nielsen:* The influence of Propranolol on the extrathyroidal metabolism. Acta Med. Scand. Suppl. **624** (1979), 31.
19. *Melmed, S., K. Nademanee, A. W. Reed, J. A. Henrickson, B. N. Singh, J. M. Hersham:* Hyperthyroxinemia with bradicardia and normal thyrotropin secretion after chronic amiodarone administration. J. Clin Endocrinol. Metab. **53** (1981) 997.
20. *Morley, J. E., S. Melmed, A. Redd, B. G. Karson, S. R. Levin, A. E. Pekary, J. M. Hersham:* Effect of vitamin A on the hypothalamo-pituitary-thyroid-axis. Amer. J. Physiol. **238** (1970), E 174.

21. *Ross, P. J., M. K. Jones, R. John:* The effect of propranolol-withdrawal on thyroid hormones in normal and hyperthyroid subjects. Clin. Endocrinol. **13** (1980), 27.
22. *Smals, A. G. H., P. W. C. Kloppenborg, W. H. L. Hoefnagels, J. I. M. Drayer:* Pituitary thyroid function in Spironolactone treated hypertensive women. Acta Endocrinol. **90** (1979), 577.
23. *Targovnik, H. M., B. E. Gluzman, A. H. Coleoni, A. Niepomniszcze:* Effects of phenylbutazone on thyroid iodine metabolism in vitro. Acta Endocrinol. **94** (1980), 64.
24. *Topolčan, O., V. Karlicek, M. Skorpilová, E. Kerndlová, A. Fikrle:* Einfluß verschiedener Medikamente auf die Plasma-Werte von Schilddrüsenhormonen. Radiobiol. Radiother. **25** (1985).
25. *Wenzel, K. W.:* Einfluß von pharmakologischen Substanzen auf die in vitro-Tests der Schilddrüsenfunktionsdiagnostik: Anlaß zu diagnostischen Irrtümern. Therapiewoche **30** (1980), 43.
26. *Wolff, J., I. L. Chaikoff:* The inhibitory action of excessive iodide upon the synthesis of diiodotyrosine and of thyroxine in the thyroid gland of the normal rat. Endocrinology **43** (1948), 174.
27. *Wolff, J., I. L. Chaikoff, R. C. Goldberg, J. R. Meier:* The temporary nature of the inhibitory action of excess iodide on organic iodine synthesis in the normal thyroid. Endocrinology **44** (1949), 504.

Chapter 20

Pharmaceutical Interaction with Radiopharmaceuticals: II. Various Labelled Compounds

Harald Deckart and *Peter H. Cox*

Interactions between drugs and radiopharmaceuticals may be caused by pharmacodynamic effect, pharmaceutical reaction or pharmacokinetic action. Examples of these three possibilities will be given in the following survey of interactions reported in the literature during the last ten years; in many cases the cause-relationship is unclear. It seems to us important to collect those observations and to give this summary of known incidence to the examining physician, who should know about these unusual findings in daily diagnostic practice.

1. Tc-99m-pertechnetate

Angiographic X-ray contrast media may lead to increased vascular permeability: penetration of blood brain-barrier was observed on the cerebral hemisphere after one-sided angiography /1/. Increased uptake in the region of cerebral ventricles during methotrexate-treatment and normalization after ceasing therapy was reported /2/. Glucocorticosteroids lead to decreased focal accumulation because of reducing perifocular edema /3/. A focal accumulation in the brain has been observed after psychotropic drugs /4/. High aluminium content of the eluate (aluminium-breakthrough, see Chapter 3 and 17) or hyperalbuminemia after ingestion of aluminium-containing antacid drugs are able to induce ^{99m}Tc-colloid-building with a change in pharmacokinetics: preferred RES-accumulation /5/. ^{99m}Tc-deposits were observed in the GI-tract due to localized bleeding caused by different drugs such as anticoagulants, antineoplastic agents, adreno-corticosteroid, indomethacin, phenylbutazone, colchicine, salicylates /12/.

2. Tc-99m-S-colloid

Hyperalbuminemia from antacid medication as well as aluminium ions in the generator eluate result in larger colloids retained in the lung capillary network /6/. Shortly after halothane narcosis a change in liver blood flow and a relative higher extraction rate in the spleen-to-liver ratio can lead to false-positive diagnosis of chronic hepatocellular damage /7/. This was also found after methylcellulose /8/, while epinephrine injected decreased spleen uptake.

3. Tc-99m-HSA-aggregates and microspheres

Iodinated antiseptics may cause a liberation from macroaggregates in vivo /9/. Pharmacodynamic effects on the microperfusion of the myocardium diagnosed with labelled microspheres can be observed after treatment with nitroglycerin and/or β-blocking agents (propranolol) /10, 11/.

4. Tc-99m-pyrophosphate

In the first week of cytostatic treatment with cyclophosphamide, doxorubin and vincristine an accumulation of labelled pyrophosphate in the kidney with special intensity has been found: a reversible toxic lesion of the tubular system with longer retention of even normally renal excreted pyrophosphate is assumed /13/.

Patients with a cardiotoxic lesion resulting from an excess treatment with adriamycine, showed a diffuse accumulation of pyrophosphate in the heart, which otherwise is known only as localized uptake in focal lesions in myocardial infarction /14/. Similar pictures have been observed in cases after a defibrillation and reanimation treatment with a diffuse accumulation of the radiopharmaceutical in the myocardium /15, 16/.

In estrogen treatment which induced gynecomastia in men, an accumulation of pyrophosphate in the breast has been found /17/. This phenomenon has not yet been reported from other drugs which may cause gynecomastia associated with estrogen excess or testosteron deficiency.

Increased calcium and phosphate serum levels lead to reduced bone uptake of pyrophosphate /18, 19/.

5. Tc-99m-diphosphonates

The influence of iron and iron-containing drugs on the retention of diphosphonates has been observed frequently. A localized muscular accumulation in the injection site of iron-dextran was reported /20/. After intravenous infusion of iron containing drugs or in cases of chronic iron overload in chronic diseases, a change in the ratio of bone-to-kidney accumulation of the renal-excreted diphosphonate complexes has been published: a decrease in bone uptake and an increase in accumulation in renal parenchyma /21, 22/. In experimental investigations a dissociation of the ^{99m}Tc-diphosphonate complex and a conversion into the renaltropic gluconate was proved in the presence of ionic iron (II) and calcium /23, 24/. An in vivo change in the complex was concluded from an increase in renal uptake even after infusion of dextrose /25/.

Long-term treatment with glucocorticosteroids results in a decrease in bone uptake /26/. This phenomenon may not be explained as an interference between pharmaceutical and radiopharmaceutical but by the drug-induced osteoporosis.

Aluminium-breakthrough in the ^{99m}Tc-eluate or treatment with aluminium-containing drugs may result in colloid building with liver and spleen uptake, which also occurs in changing pH above 8.5 /27–29/.
The identification of bone metastases with ^{99m}Tc-diphosphonate complexes can become difficult during cytostatic treatment: a weak accumulation of the

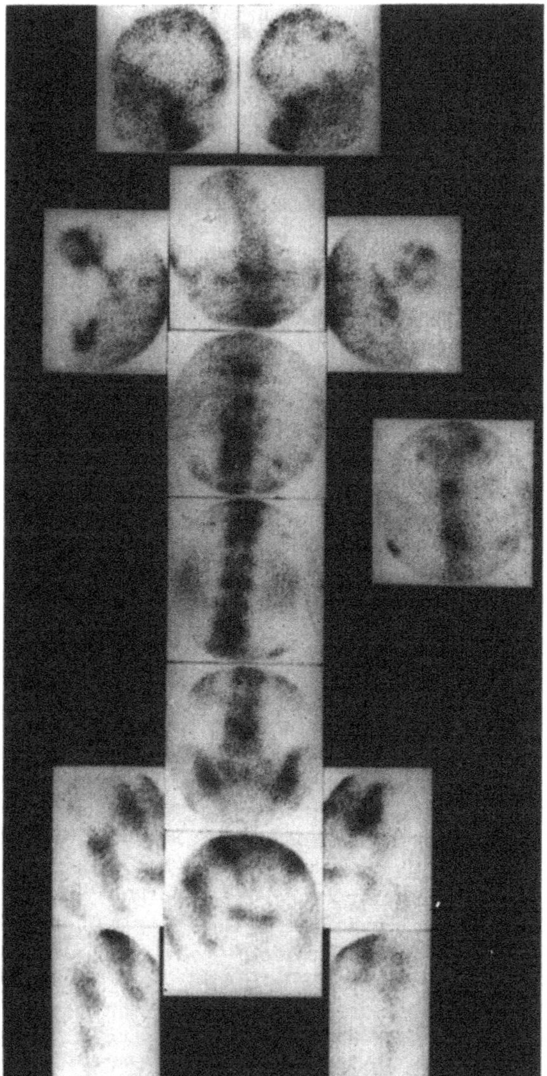

Fig. 20.1. Skeletal scintigram showing multiple metastases from breast carcinoma

radiopharmaceutical in bone lesion can be caused by the cytostatic effect on tumour tissue. Creutzig and Dach found a localized increase of uptake in the skull, containing no tumour which is characterized as "sickle-sign" because of its shape /30/. Both observations may give rise to a false-negative of false-positive diagnostic decision.

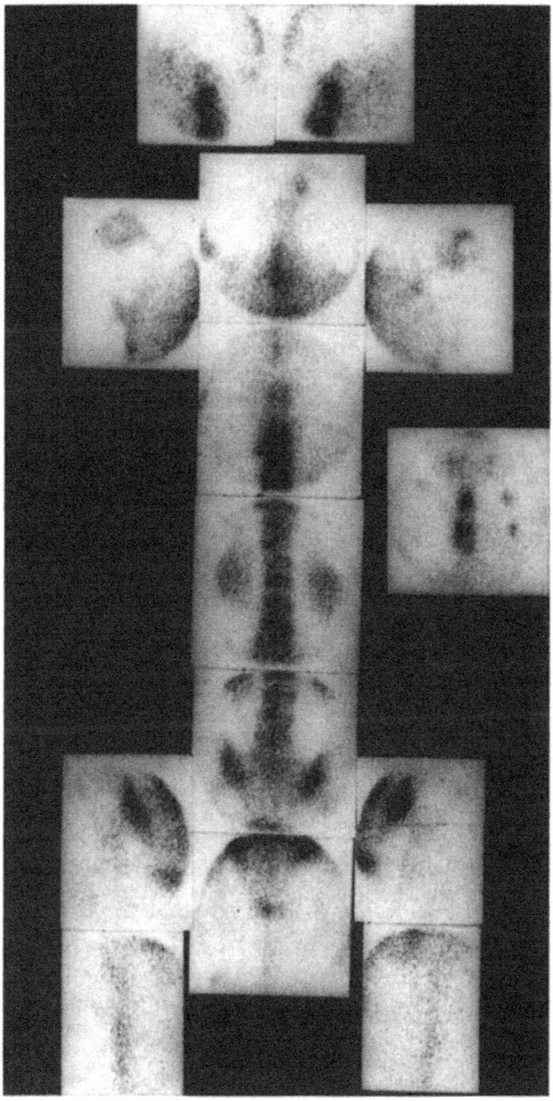

Fig. 20.2. Skeletal scintigram of the same patient following steroid therapy showing suppressed lesion uptake plus new lesions

A change in bone metabolism was shown in an observation after regional perfusion of a limb with antibodies (mephalon and actinomycin D): shortly after the bone of this limb showed an increase in the uptake of diphosphonate /31/.

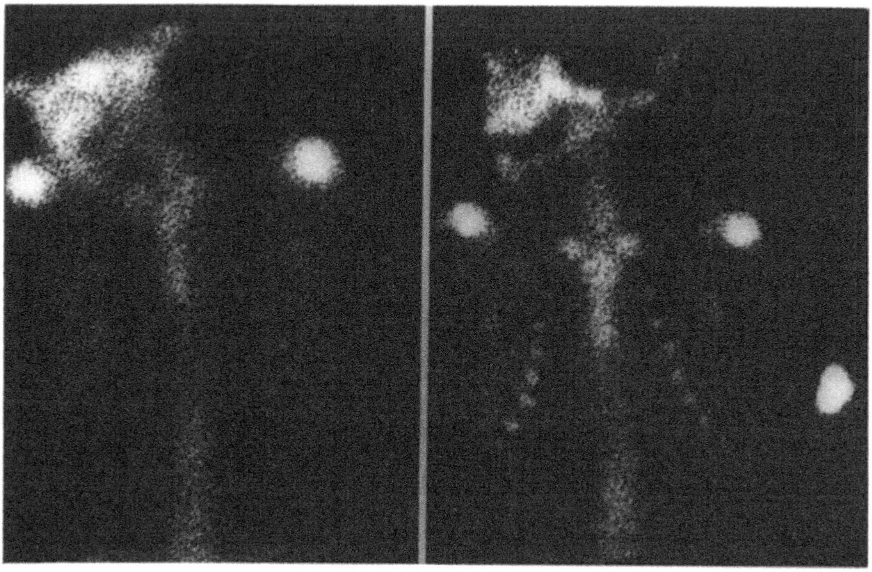

Fig. 20.3. Normal uptake of ^{99m}Tc-MDP in a child (right), depressed bone, joint and growth area uptake of ^{99m}Tc-MDP in the same patient following chemotherapy with Vincristine in a case of Ewing sarcoma (left)

6. I-131-sodium hippurate

Diuretics as furosemide, thiazole can lead to a prolonged excretion of hippurate and may cause a false-positive interpreted renogram curve. The competitive inhibition of tubular secretion by tubular load with the anti-gout drug probenecid is known, which was frequently used to investigate the excretion mechanism of newly synthesized renal-excreted radiopharmaceuticals. The similar-acting paraaminohippuric acid (PAH) was formerly used to detect latent tubular insufficiency in the form of PAH-preloading before performing a second renogram. All these drugs may induce a prolonged renal excretion pattern.

7. I-131-(or Se-75-) iodocholesterol

Drugs interacting with the hypothalamus-pituitary-suprarenal system as ACTH (which stimulates cholesterol uptake) or glucocorticosteroids (dexamethasone, hydrocortisone, prednisone, prednisolone) with their ACTH-suppressive effect may lead to weak or negligible accumulation of the organ-specific radiopharmaceutical in adrenal tissue. This pretreatment before scanning has to be taken into consideration to avoid diagnostic errors. In defined dosage and terms they are used for functional differentiation in adrenal imaging (see Chapter 4).

8. In-111-DTPA

Acetazolamide may give a false positive cisternogram by a decrease in cerebrospinal fluid production due to carbonic anhydrase inhibition /32/.

9. Se-75-Selenomethionine

This radiopharmaceutical has nowadays become obsolete for pancreas imaging because of the high radiation dose and more effective diagnostic procedures now available, such as ultrasound and CT. Interactions by growth hormone (STH), l-dopa and apomorphine treatment with false-negative interpretation and by somatostatin with the possibility of false-positive decision in pancreas imaging were published in 1979 /33/.

10. Tl-201-chloride

The normally homogenous uptake and distribution in the myocardium of the left ventricle has been shown to become diminished after treatment with propranolol, isoproterenol and cardiac digitalis glycosides /34/.

11. Xenon-133-gas

Pulmonary-ventilation studies can be influenced by drugs, acting on the pathways of the gas to the alveoli. Bronchoconstricting or broncholytic therapeuticals administered before examination can simulate localized ventilation/perfusion defect or gaseous retention in originally diseased areas (false-negative results). On the other hand, they can be used for functional investigations in lung imaging, e.g. perfusion-ventilation studies before und during shortly applied broncholytic medications.

12. Ga-67-citrate

In patients treated with bleomycine or after BCG-scarification an abnormal accumulation of ^{67}Ga-citrate in lungs has been found /35, 43/ due to a drug-induced interstitial pneumonitis and later fibrosis /36/. These false-positive results can be expected in future in further patients receiving various, especially cytostatic drugs, which are known to induce pneumonitis and interstitial lung fibrosis.

A change in normal whole-body distribution pattern of ^{67}Ga-citrate was experimentally proved and clinically observed after vincristine, cis-platine and inactive gallium citrate treatment. The latter induces a competition between radioactive and inactive gallium resulting in higher bone and liver ^{67}Ga-uptake /37, 38/. In some patients under cis-platinum treatment a ^{67}Ga-accumulation in the stomach and in the kidneys, which is not induced by a tumour, and a higher blood pool concentration with lower liver uptake have been found /43/.

Glucocorticosteroids are observed to decrease the diagnostically useful accumulation in brain tumours and in lung sarcoidosis /39, 40/. Estrogen-containing drugs lead to gynecomastia in man, which accumulates ^{67}Ga-citrate just as in phenothiazine-induced galactorrhoea /41, 42/.

An enhanced liver uptake of ^{67}Ga-citrate has been detected after ingestion of phenobarbital and iron-dextran /44, 45/.

^{66}Ga-citrate imaging of the lymphatic system may be disturbed after Lipoidol-lymphography. This iodine containing oily contrast agent is retained in the lymphatic vessels and lymph nodes over a long period (for months up to years). This blockage can induce a decrease in ^{67}Ga-citrate uptake in the diseased lymph nodes with false-negative interpretation. On the other hand, lymphadenitis, chemical pneumonitis and oil-emboli are known complications of diagnostic Lipoidol lymphography which can induce an increase in ^{63}Ga-citrate accumulation /46/.

Reviewing the above summary of published interactions, the following three types of interference have been presented: In *pharmacodynamic interaction* an interference between two pharmaceuticals occurs in their reaction on organ function: these effects are rare in nuclear medicine because of the pharmacodynamic ineffectiveness of carrier-free radiopharmaceuticals. The competitive reaction of gallium citrate and ^{67}Gallium-citrate, or of para-aminohippuric acid or probenecid with the likewise tubular-secreted ^{131}I-orthoiodo hippurate sodium are examples of pharmacodynamic interactions. The interaction of aluminium ions on ^{99m}Tc-pertechnetate with its colloid-building may be an example of *pharmaceutical interaction.* Most of the reviewed interferences have to be classified as *pharmacokinetic interactions,* in which the physiologic distribution pattern of a radiopharmaceutical is changed in the presence of a drug previously applied: a decrease or increase in radiopharmaceutical-organ-uptake by drugs, which are present or acting on the individual organ: iron changes the accumulation ratio bone-kidney of pyrophosphate; ACTH stimulates iodocholesterol uptake in the adrenal cortex, etc.

13. In this connection some additional remarks are necessary: there are

further interactions on radiopharmaceutical distribution, which are clinically important but do not touch on the subject of this Chapter.
Further *therapeutic procedures* such as surgery, radiation treatment, hemodialysis and defibrillation are influencing the normal distribution pattern: in the region of *surgical lesion* cut vessels and capillary network are due to the permeation of radiopharmaceuticals and their retention around the lesioned tissue: the increase in uptake of ^{99m}Tc-pertechnetate or further labelled compounds is well known. A higher metabolic rate in the surrounded reactive bone after fracture or surgical bone section leads to an increased uptake of osteotropic radiopharmaceuticals.
Local *radiation treatment* was due to an increased uptake of ^{99m}Tc-pyrophosphate or ^{67}Ga-citrate in extraosseous soft tissue /47–49/. In 1974 Peter Cox described further abnormalities including a diminution in physiological uptake caused by radiation effecting a suppression of osteoblastics in 1974 /50/. After *hemodialysis* an increased bone uptake of ^{67}Ga-citrate with altered biodistribution was observed /43/. The *defibrillation* reaction on heterotopic pyrophosphate uptake in the myocardium has already been considered above.

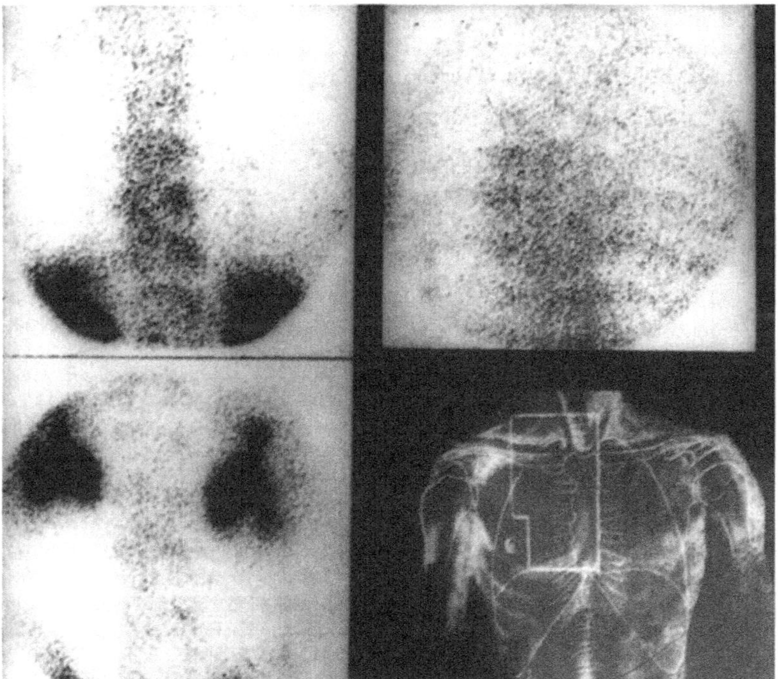

Fig. 20.4. Reduced skeletal uptake of ^{99m}Tc-polyphosphate in the radiation field as long-term effect of radiotherapy

14. Furthermore, the *atypical accumulation* of radiopharmaceuticals due to *pathomorphological* and/or *pathophysiological alterations* in several *diseases* has also to be mentioned in this context, though it does not belong to the pharmaceutical-radiopharmaceutical interactions. But it has to be taken into account in diagnostic decision and has to be known by the nuclear medicine specialist. Various examples of extraousseous accumulation of osteotropic radiopharmaceuticals may highlight the importance of these remarks: Pyrophosphates or diphosphonate complexes were found in infarcted area of the brain /51/, in brain tumour /52/, in breast carcinoma /52–54/, in the lungs in cases of breast carcinoma and parathyroid adenomas /55, 56/, in neurofibroma and neuroblastoma /57, 58/, in ovarian cancer /62/, in tumour-induced pericardial, pleural and peritoneal effusions /59–61/, in inflammation, abscesses, in degenerative changes of muscle tissue /63/, in the spleen in sickle cell disease, Thalassemia major and lymphogranulomatosis /64, 65/, in localized calcifications in soft tissue metastases, calcific valvular heart disease /66–68/, in amyloidosis /69, 70/ and showing elevated blood pool concentration in hypercalcemia just as increased serum iron levels /71/.
^{99m}Tc-S-colloid was observed to be concentrated heterotopically in kidneys in congestive heart failure /72, 73/ and in lungs in hemophilia, anemia, ileitis terminalis Crohn, breast carcinoma, falciparium malaria /74–77/. The reported cases of lung retention in liver cirrhosis may be due to shunting in the advanced state of disease.
^{133}Xenon-gas has been abnormally retained in the liver in the case of hyperlipidemia, hypercholesterinemia and steatosis hepatis /78/.

References

To ensure full documentation of every single report the shortest possible bibliographical data were used.

1. Radiology **92** (1969), 1467.–2. Clin. Nucl. Med. **3** (1978), 22.–3. JAMA **232** (1975), 1261.–4. J. Nucl. Med. **17** (1976), 527.–5. J. Nucl. Med. **19** (1978), 381.–6. J. Nucl. Med. **15** (1974), 1220.–7. Radiology **118** (1976), 303.–8. J. Nucl. Med. **13** (1972), 211.–9. J. Nucl. Med. **18** (1977), 1139.–10. Circulat. Res. **28** (1971), 263.–11. J. Nucl. Med. **17** (1976), 539.–12. U. S. Pharmacist **3** (1978), 35.–13. Radiology **128** (1978), 165.–14. J. Nucl. Med. 18 (1977), 680.–15. Clin. Nucl. Med. **2** (1977), 6.–16. Circulation **54** (1976), 980.–17. Clin. Nucl. Med. **2** (1977), 206.–18. Life Sci. **22** (1978), 1261.–19. J. Nucl. Med. **22** (1981), 526.–20. J. Nucl. Med. **16** (1976), 374.–23. J. Nucl. Med. **17** (1976), 208.–24. Radiology **140** (1981), 197.–25. Brit. J. Pharm. Pract. 1980, 17.–26. Semin. Nucl. Med. **12** (1982), 184.–27. Intern. J. Nucl. Med. Biol. **3** (1976), 37.–28. Radiology **126** (1978), 813.–29. N. Engl. J. Med. **196** (1977), 1389.–30. Eur. J. Nucl. Med. **6** (1981), 99.–31. Clin. Nucl. Med. **2** (1977), 451.–32. J. Nucl. Med. **19** (1978), 501.–33. J. Nucl. Med. **20** (1979), 543.–34. J. Nucl. Med. **17** (1976), 537.–35. Cancer **36** (1975), 1966.–36. Amer. J. Clin. Pathol. **58** (1972), 501.–37. Radiology **117** (1975), 709.–38. Proc. Amer. Soc. Clin. Oncol. **19** (1978), 197.–39. J. Nucl. Med. **18** (1977), 617.–40. Amer. J. Roentgenstr. **122** (1974), 744.–41. Radiology **124** (1977), 169.–42. J. Nucl. Med. **19** (1978), 560.–43. Semin. Nucl. Med. **9** (1979), 131.–44. J. Nucl. Med. **19** (1978), 1089.–45. J. Nucl. Med. 17 (1976), 456.–46. J. Nucl. Med. **16** (1975), 374.–47. Clin. Nucl. Med. **2** (1977), 93.–48. J. Nucl. Med. **19** (1978), 225.–49. Thorax

27 (1972), 543.–50. Brit. J. Radiol. **47** (1971), 1816.–51. J. Nucl. Med. **15** (1974), 207.–52. J. Nucl. Med. **15** (1974), 1054.–53. Radiology **109** (1973), 393.–53. Radiology **112** (1974), 655.–55. J. Nucl. Med. **18** (1977), 890.–56. J. Nucl. Med. **22** (1981), 888.–57. J. Nucl. Med. **15** (1974), 1207.–58. Radiology **111** (1974), 127.–59. J. Nucl. Med. **20** (1979), 1102.–60. J. Nucl. Med. **17** (1976), 29.–61. Radiology **139** (1981), 699.–62. J. Nucl. Med. **17** (1976), 29.–63. Clin. Nucl. Med. **8** (1983), 165.–64. J. Nucl. Med. 17 (1976), 108.–65. J. Nucl. Med. **19** (1978), 976.–66. Brit. J. Radiol. **48** (1975), 724.–67. J. Nucl. Med. **19** (1978), 974.–68. J. Nucl. Med. **18** (1977), 776.–69. J. Nucl. Med. **16** (1975), 238.–70. J. Nucl. Med. **18** (1977), 1086.–71. J. Nucl. Med. **17** (1976), 208.–72. J. Nucl. Med. **15** (1974), 564.–73. Radiology **127** (1978), 185.–74. J. Nucl. Med. **14** (1973), 687.–75. J. Nucl. Med. **17** (1976), 794.–76. J. Nucl. Med. **14** (1973), 201.–77. J. Nucl. Med. **15** (1974), 1028.–78. J. Nucl. Med. **20** (1979), 450.

Chapter 21

Radiopharmaceuticals Toxicity and Safety

Peter H. Cox

1. Introduction

Any substance which is intended for administration to humans or animals for medicinal purposes, is required to meet certain criteria with regard to safety in use. This is also true for radiopharmaceuticals whether they be for diagnostic purposes or for radiotherapy.

Toxicity tests can be classified into two basic groups, acute toxicity studies to establish safe dose levels for single dose administration and chronic to evaluate the effects of prolonged use. Within this context it also has to be possible to evaluate the late effects of a single dose administration. This latter aspect is particularly relevant to radiopharmaceuticals with respect to radiation induced damage.

The established testing procedures were designed to evaluate drugs used for therapeutic purposes and there is genuine doubt as to their suitability for testing radiopharmaceuticals.

As a general rule it is true to say that the main risk with radiopharmaceuticals is that of radiation induced damage. The labelled compounds are only present in miniscule chemical concentrations so that the question of chemical toxicity seldom arises, although this is not always the case as we will observe later.

Regulatory authorities have tended to adopt the approach that radiopharmaceuticals should be required to pass the same safety tests as therapeutic drugs. The high specific activity of most radiopharmaceuticals makes it practically impossible to provide them in concentrations suitable to carry out LD 50 and other tests in a realistic manner.

There is also a real risk that the chemical form of a radiopharmaceutical complex may change when concentrations are raised in this way thus rendering any study irrelevant. These criteria alone provide a strong case for a specific set of safety tests for radiopharmaceuticals. Let us therefore consider the possibilities with respect to known facts and in relation to existing safety testing.

2. Radiation dose

The most likely form of biological damage to the patient is radiation induced damage. To minimalise the risk the minimum dose which produces clinical useful data has to be related to the radiation dose to the target organ, i.e.

the organ receiving the highest radiation dose. In this respect the bladder and intestines should not be forgotten. Biodistribution studies in normal animals and in pathophysiological models can be helpful in predicting human radiation doses but the possibility of interspecies variation remains a constant and insoluble problem.

Biological damage caused by radiation can be considered under several headings: direct intracellular damage causing disturbances in organ function and/or cellular death, carcinogenicity and genetic damage, mutagenicity.

3. Intracellular damage

A number of radionuclides used in vivo diagnosis, decay by electron capture or internal conversion which may be accompanied by the release of low energy auger electrons with a very short range of tissue penetration. In such cases the site of localisation of the radionuclide within the cell will greatly influence the type and degree of damage produced. Decay within the nucleus may produce a radiotoxic response a factor 100 or more times greater than that associated with the cytoplasm or cell membrane. Hofer /1/ has discussed this in relation to the use of radiolabelled blood cells and Dandanudi /2/ has suggested that ^{201}Thallium may cause more damage to the gonads through its auger electrons, than has been supposed. This latter study in particular emphasises the need for adequate evaluation of the potential of a radiopharmaceutical to cause genetic damage.

4. Genetic damage and mutagenicity

Thakur and McAffee /3/ have recently reviewed the significance of chromosomal aberrations induced in ^{111}Indium labelled leucocytes and suggested that it is desirable to develop alternative reagents which bind to the cell membrane or cytoplasm only, thereby reducing the radiation dose to the nucleus.

Phagocytes labelled with radiocolloid /4/ are also more rapidly eliminated from the circulation than normally and although this may be a normal clearance pattern of such cells the possibility that it may be due to radiation damage has not been eliminated. In the latter case cells surviving but with nuclear damage become a real possibility.

This problem is not only associated with cells which have been labelled by direct methods but may be incidental to the injection of radiopharmaceuticals. The potential effect of Thallium on the gonads has already been mentioned and Stevenson et al. /5/ have reported chromosomal damage to erythrocytes from the intraarticular injection of ^{198}Gold colloid. Gumpel /6/ made similar observations concerning intraarticular ^{90}Y colloid producing chromosome damage in lymphocytes.

Such problems are in fact not new, More and Shastry /7/ have indeed sug-

gested that naturally occuring potassium – 40 present in intracellular fluid plays an important role as a natural mutagenic agent.
Chromosome aberrations are relatively easy to detect because they are visible microscopically in the prometaphase stage of cell division. Mutagenicity is however more difficult to detect because the effects may skip a generation. A way to surmount this problem is to use the Ames test for mutagenicity /8/. This test uses salmonella typhimurium as test organism which has a mutant gene for the enzyme phosphoribosyl adenosine triphosphate (ATP) synthetase. This enzyme is required for histidine synthesis and the bacteria is unable to grow in a histidine deficient medium unless a reverse mutation is induced. Because many chemicals are not mutagenic or carcinogenic unless metabolised by the endoplasmic reticulum rat liver microsomes are added to the medium. The test is both rapid and sensitive but the results are considered to be contraversial in direct relation to man.

5. Carcinogenicity

Induced carcinogenicity may have a very long latent period and for this reason alone may be difficult to detect in animal studies at all. At the chemical and radiation dose levels we are considering the incidence whilst perhaps present may not be detectable. For instance an 0.01 incidence of cancer is equivalent to 20,000 people per 200 million head of population. Animal tests to detect this level of incidence would require 30,000 animals. The extrapolation of animal studies is thus a crucial question /9/. This remains a difficult problem but fortunately the incidence of radiation induced cancer attributable to radiopharmaceuticals is low and the statistics from human data is favourable and only related to therapeutic dose levels /10/.

6. Transient radiation effects

The incidence of transient side effects due to radiation from radiopharmaceuticals is exclusively related to therapeutic doses. Sialadenitis has been observed to develop in 11 % of patients who had been treated for thyroid cancer with ^{131}Iodine /11/.
In patients treated with intraarticular ^{198}Gold for rheumatoid synovitis erythema of the skin at the injection site coupled with swelling and joint pain /12/ has been reported. Such effects, together with carcinogenicity are unlikely to be observed in animal studies partly because there are no suitable animal models to produce a comparable biodistribution to the diseased human state. The only adequate method available is a good epidemiological record of the follow up of treated patients.

7. Chemical toxicity

Whilst it is by now abundantly clear that radiation damage is the greatest toxic risk presented by radiopharmaceuticals the question of chemical toxicity is relevant with reference to carrier molecules, exipients and stabilisers especially with respect to labelling kits. Metallic ions of Arsenic, Antimony and Tellurium have been shown to cause chromosome damage in cell cultures, although at concentration levels far above the levels at which they are present in radionuclide complexes for in vivo use. Bleomycin /13/ effects lung tissue and bone marrow but only upon prolonged administration. Selenomethionine (non radioactive) has been reported to interfere with crossover distribution along the X chromosome of Drosophila /14/.

The majority of radionuclide complexes used in radiopharmacy are present at homeopathic levels of concentration so that chemical toxicity data is almost irrelevant. Gallium as nitrate inhibits cell growth in culture at 10 ng/ml concentration levels and produces cell death at 50 ng/ml /15/. ^{67}Gallium citrate as used for in vivo diagnosis contains approximately 0.002 μG per ml which means that an average patient dose contains about 0.006 μG. In the patient after injection this represents a concentration of around 0.001 μG/litre which is a dilution factor of approximately 1×10^6 below the level known to effect cell growth.

It should not be forgotten however, that as the level of radioactivity falls on storage, the proportion of carrier present rises and whilst this may once again never reach a toxic level it may affect biodistribution which is an unwanted side affect. ^{197}Mercury or ^{203}Mercury Chlormerodrin for instance may induce diuresis when used near the end of its useful storage life /16/.

When considering radiopharmaceutical toxicity the exipients in the formulation may be more relevant than the radionuclide. Benzyl alcohol 0.9 % has been widely used as a preservative and is 60 % more toxic than sodium pertechnetate.

8. Technetium labelling kits

Acute toxicity studies on the components of some technetium labelling kits (unlabelled) have been reported. Fisher reported that up to 350 mg/kg of stannous ions produced no histopathological changes in rabbits /17/. This dose level is 35,000 times higher than the maximum dose of stannous ions likely to be injected into man. The LD 50 of pyrophosphate and polyphosphate ions in rats following intravenous administration were 41 and 29 mg/kg respectively, although changes in serum calcium concentration were observed at dose levels starting at 20 mg/kg /18/. Nevertheless this bears no relationship to human diagnostic dose levels.

9. The relevance of acute toxicity testing to radiopharmaceuticals

The most common acute toxicity test is the LD 50 in which groups of animals are given increasing concentrations of the test substance, as a single dose, to determine a dose level at which 50 % of the animals die. Although still official the test has been sharply criticised in recent years. In such a test 4 species of animal are required to be tested of which one in a non rodent. The most commonly used are mice, rats, rabbits and dogs.

Some idea of the relevance of this test to radiopharmaceuticals has been given already above, but let us consider a specific example to illustrate the technical problems involved.

If gallium is present in the diagnostic injection of ^{67}gallium-citrate at a concentration of 0.002 μg/ml and the chemical cell concentration known to be lethal, from cell culture studies, is 50 μG/ml it would require a dose of 25,000 ml to produce 50 μG of gallium citrate and about 200 times that would be required to kill a 250 G rat. The 50 μG quantity of gallium alone represents about 25 Curies of injectable material. Such a test clearly cannot be carried out.

For the same reasons and because of the short halfe lives of the radionuclides used chronic 90 day and six month tests are also irrelevant.

It is possible to conclude that a specific protocol for the safety testing of radiopharmaceuticals could be developed which should include tests for:

1. Radiation damage (acute).
2. Radiation damage (long term).
   Chromosome damage
   Mutagenic effects
3. Allergenic response to single and repeat dosage.

With the exception of organ function studies much of this data can be obtained using in vitro cell culture techniques to reduce the burden of in vivo studies.

A further important factor in determining the safety of radiopharmaceuticals is the accurate reporting of observed side effects in patients and a comprehensive follow up particularly of therapy patients.

References

1. *Hofer, K. G.:* Microdosimetry of labelled cells. In: Blood cells in Nuclear Medicine. Vol. II Ed *G. F. Feuger.* Martinus Nijhoff, Boston, The Hague (1984) 224.
2. *Dandanudi, V. R.,* et al.: Radiotoxicity of Thallium-201 in mouse testes. J. Nucl. Med. **24** (1983) 145.
3. *Thakur, M. L., McAffee, J. G.:* The significance of chromosomal abberation in 111-Indium labelled lymphocytes. J. Nucl. Med. **25** (1984), 922.
4. *Oberhausen, E., J. J. Schroth:* Phagocytic labelling of migratory blood cells and its clinical applications. In: Blood cells in Nuclear Medicine Vol. II. Ed *G. F. Feuger.* Martinus Nijhoff, Boston, The Hague (1984), 199.
5. *Stevenson, A. C., et al.:* Chromosome damage in patients who have had intra articular injections of radioactive gold. Lancet **1** (1971) 837.

6. *Gumpel, J. M.:* The role of radiocolloids in the treatment of Arthritis Rheumatology and Rehabilitation XIII (1974), 1.
7. *Moore, F. D., Sastry, K. S. R.:* Intracellular Potassium-40 K as a primordial gene irradiator. Proc. Nat. Acad. Sci USA. **79** (1982), 3556.
8. *Ames, B. N., et al.:* Methods for detecting carcinogens and mutagens with the salmonella/mammalian microsome mutagenicity test. Mutat Res. **31** (1975), 347.
9. *Klaasen, C. D.:* In: The Pharmacological Basis of Therapeutics Ed. *Goodman and Gilman.* Macmillan New York, London. 6th Ed (1975), 1606.
10. *Tubiana, M.:* Risks associated with the use of radiopharmaceuticals. In: Nuclear Medicine and Biology IV Ed *C. Raynaud.* Pergamen Press Oxford (1982), 2603.
11. *Allweis, P., et al.:* Sialadenitis following I—131 therapy for thyroid carcinoma. J. Nucl. Med. **25** (1984), 755.
12. *Fine, P. H., et al.:* Intraarticular radioactive gold in the treatment of rheumatoid arthritis of the knee. Arth. & Rheum. **10** (1967), 278.
13. *Borstein, R. S., et al.:* Cytogenic effect of bleomycin therapy in man. Canc. Res. **31** (1971), 2004.
14. Editorial comment: Nature **304** (1983), 286.
15. *Rasey, J. S., et al.:* Tumour cell toxicity of stable gallium nitrate enhancement by transferrin and protection by iron. Eur. J. Canc. Clin. Oncol. **18** (1982), 661.
16. *Miller, J. P.:* Radiopharmaceutical toxicology and dosimetry. Nucl. Med. **8.** Supplement.
17. *Fischer, H. W.:* Colloidal stannic oxide. Animal studies on a new hepatobiliary agent. Radiology. **68** (1957), 488.
18. *Stevenson, J. S.* et al.: The toxicity of Sn-pyrophosphate. Clinical manifestations prior to acute LD 50. J. Nucl. Med. **15** (1974), 252.

Chapter 22

Absorbed Dose Calculations for Biologically Distributed Radiopharmaceuticals

Manfred Tautz

1. Fundamentals

The growing application of radioisotopes to diagnosis and therapy results in increasing radiation doses for the individual examined as well as for the whole population.
Generally, the internal radiation absorbed dose depends on biological parameters, the distribution and kinetics of a radiopharmaceutical as well as on the physical data of the corresponding radioisotopes.
Physicists working in X-ray departments or in radiation therapy use measuring instruments to solve dosimetric problems. Up to now there have not been any dosimeters to measure the internal radiation dose in the field of incorporation dosimetry.
Therefore, the radiation dose has to be calculated for different organs after application of radiopharmaceuticals.
The historically latest concept for calculating the internal radiation dose results from work carried out by the members of the Medical Internal Radiation Dose Committee (called MIRD).
According to the new formalism, the absorbed dose from radiopharmaceuticals can be calculated with the following general equation /1/:

$$\bar{D}\,(\tau \leftarrow \sigma) = \frac{\tilde{A}}{m_\tau}\,\Sigma_i\,\Delta_i\,\Phi_i \tag{1}$$

where
D = mean value for radiation dose
τ = aim volume (target) of source radiation σ
m = mass of target organ

The first equation is valid for the special case, when target and source have the same volumes:

$$\bar{D}\,(\tau \leftrightarrow \tau) = \frac{\tilde{A}}{m_\tau}\,\Sigma_i\,\Delta_i\,\Phi_i \tag{2}$$

with
$\tilde{A}$ = A tilde over the symbol for activity denotes a time integral

$$\tilde{A} = \int A dt \tag{3}$$

as the cumulated activity.
In another form:

$$\frac{\tilde{A}}{m_\tau} = \int C\,(t)\,dt \tag{4}$$

This is the time integral of radioactivity concentration in the target organ. This quantity, which is called retention function, can be measured with a clinical whole-body counter.

The mathematical form is

$$C(t) = \int_0^\infty C_{0i}\, e^{-0.693 \frac{t}{Teff_i}} \tag{5}$$

$$C(t) = \Sigma\, C_i(o)\, e^{-0.693 \frac{t}{Teff_i}} \tag{6}$$

After total decay:

$$C(\infty) = 1.44 \cdot C_i(o) \cdot Teff_i \tag{7}$$

with

$$T_{eff} = \frac{T_p \cdot T_b}{T_p + T_b} \tag{8}$$

By means of graphic evaluation it is possible to determine the biokinetic data which are necessary for radiation dose calculation, namely:

- the biological half-life of intake and excretion;
- the percentage of single components of the retention function;
- the portion of the totally administered radioactivity deposited in an organ extrapolated to the time of administration of the radionuclide.

Further symbols in equation (1):

Δ_i = Mean energy emitted per unit cumulated activity for i^{th} radiation component of the decay scheme with high- and low-energy photons, beta particles, conversions electrons and Auger electrons.
It is necessary to know the actual decay scheme of the radioisotopes. Then the mean energy per unit cumulated activity is /1/:

$$\Delta\left(\frac{g \cdot rd}{\mu Ci \cdot h}\right) = n\left(\frac{particles}{nuclear\ transf.}\right) \times 1.332 \cdot 10^8 \left(\frac{nuclear\ transf.}{\mu Ci \cdot h}\right)$$
$$\times E\left(\frac{MeV}{particle}\right) \times 1.602 \cdot 10^{-8} \left(\frac{g \cdot rd}{MeV}\right) = 2,13 \cdot n \cdot E$$

In the SI:

$$\Delta\left(\frac{kg \cdot Gy}{Bq \cdot s}\right) = n\left(\frac{particle}{nuclear\ transf.}\right) \times E\left(\frac{J}{particle}\right) = n \cdot E$$

Φ_i = Absorbed fraction defined as energy from the ith radiation component in source absorbed in target volumen divided by energy of i th radiation component emitted by the source region. It is assumed that both the source and target are regions in a homogeneously absorbing material and the activity is uniformly distributed in the source region. In practice one finds a uniform distribution of activity only in rare cases. This is a source of errors.
If the energy is largely absorbed in a single target and only a small fraction of the energy is absorbed, the absorbed fraction Φ is relati-

vely insensitive to alterations in both the size and the shape of source and target. The absorbed fraction will approach to the unity at low energies.

Generally, the absorbed fractions can be found in the tables of MIRD for photon sources which are uniformly distributed in various organs of a heterogeneous human phantom /2, 3/ Figure 22.1. The total-body phantom consists of three principal sections (reference man).

a) An elliptical cylinder representing the arms, torso and hip;
b) an elliptical cone representing both the legs and the feet;
c) an elliptical cylinder representing both the head and neck.

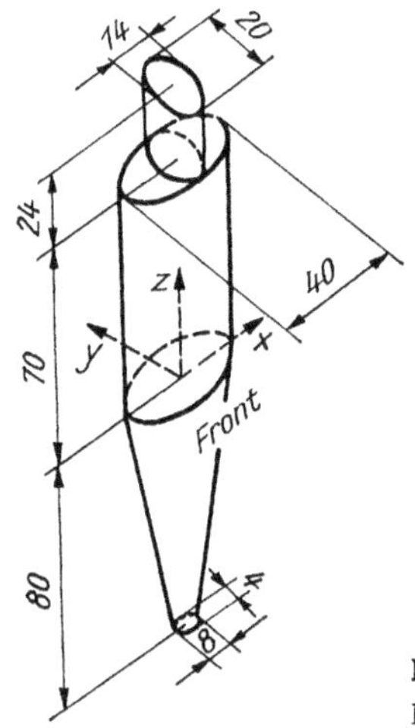

Fig. 22.1. MIRD-Reference man. The adult human phantom (dimensions in centimeters)

This adult human phantom is valid for a standard man (m = 70 kg). Furthermore, there are mathematical descriptions for the single organs. These "mathematical" organs approximate the major features of size, shape, mass and location of the real organs.

For these different phantom organs there are tables including the quantity Φ as a function of the photon energy /3/.

An example for thyroid gland is shown in Table 22.1.

Table 22.1. Parameter Φ for source in thyroid and some target organs for discrete energies

	0.05	0.1	0.2	0.5	1.0 MeV
Kidneys	—	0.2 E–4	0.3 E–4	0.65 E–4	0.99 E–4
Liver	0.2 E–3	0.6 E–3	0.7 E–3	0.94 E–3	0.12 E–2
Lungs	0.37 E–2	0.37 E–2	0.34 E–2	0.4 E–2	0.4 E–2
Thyroid	0.48 E–1	0.3 E–1	0.3 E–1	0.3 E–1	0.3 E–1
Total body	0.48	0.3	0.3	0.3	0.3

It is very important that the fundamental equation (1) for the radiation dose is quite general and not restricted to any particular kind of radiation. This formalism is independent of the specific gamma-ray constant and of the average geometrical factor for the absorber in an earlier mathematical formalism /4/. In practice, however, the classification of radiation in two groups —non-penetrating and penetrating—is very useful /1/.

The term non-penetrating is applied to those radiation components, for which only an insignificant amount of energy is absorbed outside a source volume. As the human body and particular organs, for which the absorbed dose is to be calculated, have dimensions which are considerably larger than the range of Beta-particles and low-energy photons with energies less than 11.3 keV, these radiation components are treated as nonpenetrating.

This energy means that 95 % of radiation will be absorbed in water in a distance of 10 mm.

Two cases are distinguished:

$\Phi\ (\tau \leftrightarrow \tau) = 1$ target equal source

$\Phi\ (\tau \leftarrow \sigma) = 0$ difference between target and source

For penetrating radiation the values for Φ as a function of radiation energy of the photons and particular organs can be borrowed from Tables /3/.

In MIRD-Pamphlet No. 11 /5/ the mean energy per unit cumulated activity (equilibrium dose constant), the mass m of the target organ and the absorbed fraction Φ are summarized. This is convenient to introduce the quantitiy

$$S\ (\tau \leftarrow \sigma) = \frac{1}{m_\tau}\ \Sigma_i\ \Delta_i\ \Phi_i \tag{10}$$

Most of the biological data needed for the estimation of dose are embodied in the cumulated activity Ã, while the remaining portion of parameters in Eq.1 involves physical and anatomic data. Therefore, S can be tabulated as defined by Eq.10 for radionuclides and source-target configurations commonly used in nuclear medicine.

The average dose from a specified radionuclide can be calculated now by the simple formula

$$\bar{D}\ (\tau \leftarrow \sigma) = \tilde{A} \cdot S\ (\tau \leftarrow \sigma) \tag{11}$$

As the MIRD scheme refers only to absorbed dose calculations for specified pairs of source and target organs, including the total body, the concept was extended to the estimation of the additional target dose resulting from radioactivity distributed within the remaining parts of the body.

Generally, the totally absorbed dose $\bar{D}_\tau$ in any target organs is composed of three parts (Figure 22.2.), where

$D_{\tau \leftrightarrow \tau}$ = dose due to self-irradiation of the target organ

$D_{\tau \leftarrow \sigma}$ = dose due to irradiation from other source organs

$D_{\tau \leftarrow RB}$ = dose due to irradiation from radioactivity in the remaining body

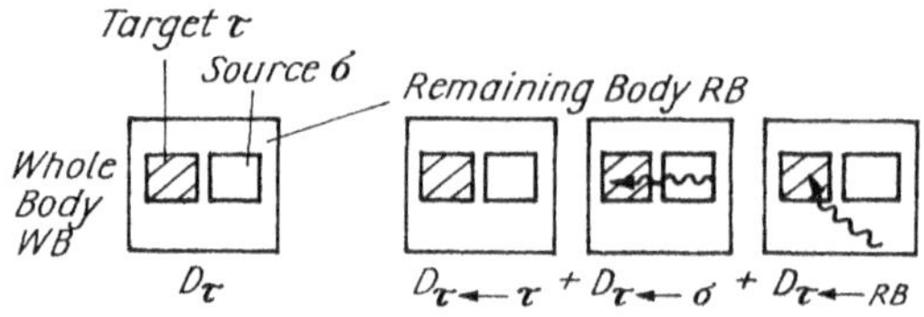

Fig. 22.2. Relation of target, source and remaining body

2. Practical problems for radiation dose calculation

a) Estimation of the age-dependent internal dose from radiopharmaceuticals
The age-dependence of the radiation dose absorbed in a target organ or in the total body is determined by the mass, the size, the shape of the organs and the distance to the other organs. A correction has to be applied to all photon energies emitted by the radionuclides. It is necessary to calculate the values of Φ for all interesting organs in age-dependence. For short-lived radionuclides, such as ^{99m}Tc generally recommended for use in children, variations of the biological half-life are of minor importance to the absorbed dose:
Therefore, the absorbed dose to infants and children may be calculated by adopting the biological half-lives of adults.
As an example of radiopharmaceuticals with age-dependent biokinetic data dose calculations were performed for ^{131}I sodium iodide (Figure 22.3.). These indicate an approximately tenfold thyroid dose to the newborn compared to the adult /6/.

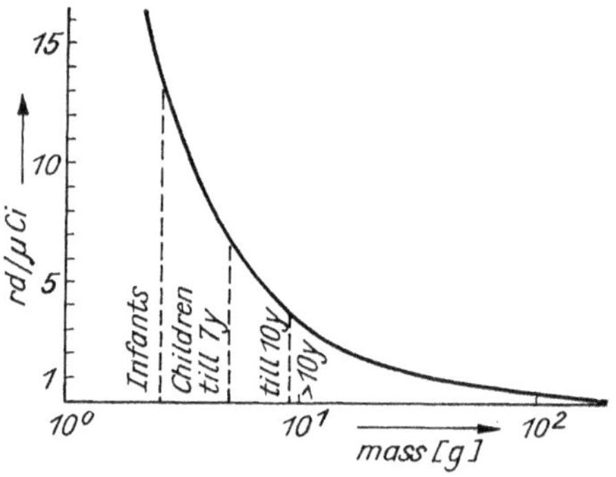

Fig. 22.3. Radiation dose as a function fo thyroid mass and age-dependence (Tautz and Ertl 1976)
Parameters: ^{131}I-Iodide
$T_{eff} = 3$ d
$U = 0{,}75$
$A = 1\ \mu$Ci

Table 22.2. Effective half-life of ^{131}I-iodide in age-dependence

Age (years)	Effective half-time (days)
0	5.1
0,5	6.0
1	6.1
3	6.3
5	6.3
10	6.8
15	6.9
adult	7.6

Table 22.2. shows how the effective half-life is varied from 5.1 days for newborns to 7.6 days for adults.
These results are very important for radiation protection purposes and for the assessment of the individual radiation risk.

b) Whole-body retention studies

Whole-body profiles for the total body (Figure 22.4.) were measured with a clinical Whole-Body Counter /8/ and with a special measuring technique for single organs (e.g. with collimators). By means of profiles whole-body-retention was constructed with some exponential components (Figure 22.5.).
On healthy persons as well as on patients with chronic pyelonephritis and with renal insufficiency of different severity whole-body-retention measurements were carried out up to 75 days after intravenous injection of ^{131}I-hippurate (Figure 22.6.). Whole-body-retention curves of persons with

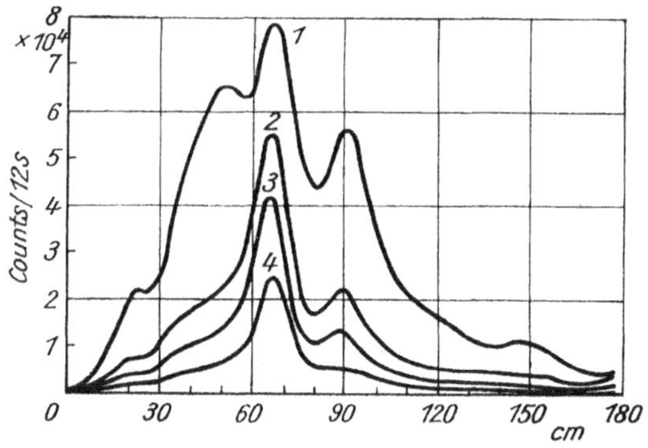

Fig. 22.4. Whole-body profile curves for Tc-99m-Thioglucose (Tautz et al. 1979)
1 = 0 h p. i
2 = 3,5 h p. i
3 = 6,5 h p. i
4 = 9,1 h p. i

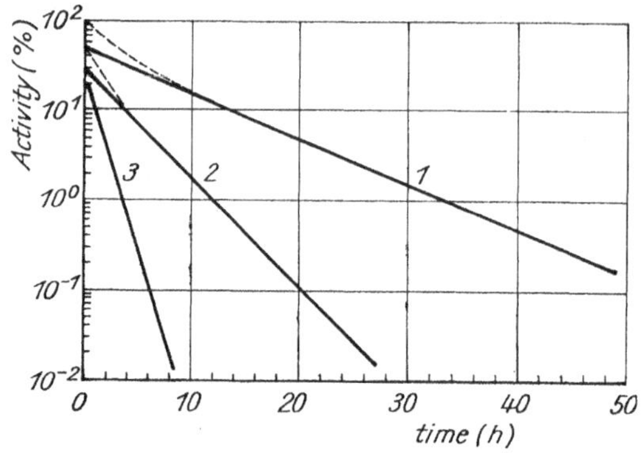

Fig. 22.5. The corresponding whole-body-retention functions for the whole-body profiles in Figure 21.4. (Tautz et al. 1979)
1 = Q_1 = 50 %;
2 = Q_2 = 28 %;
3 = Q_3 = 22 %;
T_{eff} = 5,8 h
T_{eff} = 2,4 h
T_{eff} = 0,8 h

healthy kidneys could be divided into two components. The first half-live (T_{B1}) is very short, i.e. it comprises 20 minutes.

For renal insufficiency there are three components. The last component has a half-life of up to 35 days and longer and an extrapolated activity of 1.5 to 10 % resulting from ^{131}I-iodide split-off in vivo (Table 22.3.). Incorporation dosimetric calculations, according to the MIRD conception, allowed to determine a whole-body dose increased by the factors of 2.5 to 40 for patients with renal insufficiency compared to those with healthy kidneys (Table 22.4.).

By in vivo splitting-off of ^{131}I-iodide in long-time retention, due to the excretion insufficiency, radiation doses were determined on the unblocked thyroid gland between 0.36 and 1.20 Sv or 36—120 rd respectively.

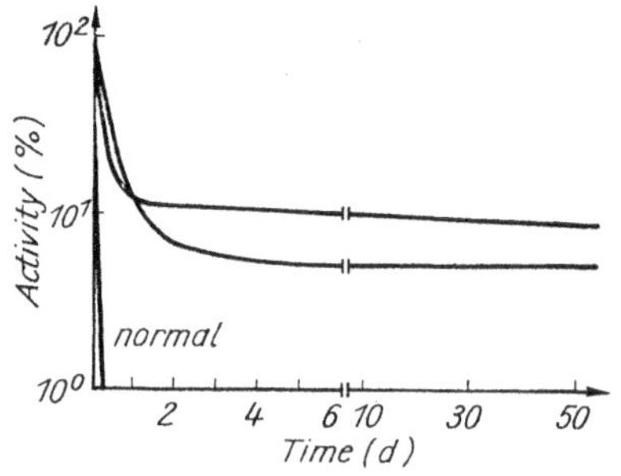

Fig. 22.6. Whole-body-retention curves of ^{131}I-hippurate

Table 21.3. Components of a whole-body retention curve (Patient: chronical Pyelonephritis)

	Components		
	I	II	III
Activity (%)	83	7	10
Biological half-life (d)	0.15	1.25	> 35
Portion of whole-body radiation dose (%)	11.5	8.5	80

Generally it should be emphasized:

The radiation dose is often a function of the severity of disease.

c) I shall now make some sumarizing remarks about radiopharmaceuticals for clinical evaluation of the disease of the thyroid gland (Table 22.5. and Table 22.6.) with regard to radiation protection results /10/:

These recommendations are made by the Bureau of Radiological Health, Food and Drug Administration, Department of Health, Eduction and Welfare, USA.

^{131}Iodine as sodium iodide:

For uptake measurements using a single crystal with an average diameter of about 2 inches (5 cm) and a thickness of at least 1 inch (2.5 cm) at a standard distance of 10 inches (25 cm) from crystal to patient.

Recommended maximum activity for uptake percentage measurement are as follows:

for adults 6 μCi,

for children generally 2 μCi.

Imaging with the scanner and/or scintillation camera is based upon the percentage of iodine uptake and estimated thyroid gland weight. An activity of 30 μCi is recommended for euthyroid patients whose glands are of normal size. An activity of 100 μCi would be considered to be an upper limit especially suitable for hypothyroid patients.

Table 22.4. Radiation dose with ^{131}I-Hippurate for the whole-body (mrd)

	Activity	References	Normal persons	Patients (Kreatinin 2–6.4 mg/ 100 ml)
^{131}I-Hippurate (Renogr.)	20 μCi	0.2	0.15–0.4	0.4– 7.7
^{131}I-Hippurate (Sequence scintigraphy)	300 μCi	3.0	2.25–6.0	6.0–116.0
Relation normal persons : patients = 1 : 2.5–40				

X-ray exposure pelvis
75–150 mrd

Table 22.5. Physical parameters of ^{131}I, ^{123}I and ^{99m}Tc

	^{123}I	^{131}I	^{99m}Tc
Physical half-life	13 h	8.1 d	6 h
$E\gamma$ (MeV)	0.028; 0.867	0.030; 0.046	0.141; 0.883
	0.159; 0.836	0.080; 0.026	0.143; 0.0003
	0.529; 0.11	0.284; 0.058	
n;		0.364; 0.820	
mean number		0.637; 0.065	
p. disintegration		0.723; 0.017	
Dose constant	0.0610	0.4085	0.0369
Δ np (g · rd/μCi · hr)			

^{123}Iodine as sodium iodide:

For uptake measurements and for imaging the same equipments can be used as for ^{131}I.
The following oral activities are recommended:

Measurement at 4–6 h: 10 μCi,
Measurement at 24 h: 20 μCi.

For imaging an activity within the range of 100–400 μCi is recommended.
^{99m}Technetium as pertechnetate:

An imaging activity of 3–10 mCi is generally adequate (the author's opinion: 1–1.5 mCi should suffice). Activities of 15–20 mCi have been used to improve visualization. At these activity levels the physician should consider the radiation dose delivered to the stomach and colon.

d) Now I am going to make some remarks about the relation between the radiation dose from X-ray examinations and nuclear medicine procedures: As early as 1965 Edward Smith wrote /11/:

Table 22.6. Typical absorbed dose due to thyroid diagnostic radiopharmaceuticals

	A	Critical organ	D_{Co} (actual mrd)	D_{Gonad} (mrd)
Uptake				
^{123}I-Iodide	10– 20 μCi	thyroid	110– 220	M: 0.1– 0.2
				F: 0.2– 0.4
^{131}I-Iodide	6 μCi	thyroid	6,600	M: 0.52
				F: 0.84
Scan				
^{99m}Tc	5– 10 mCi	thyroid	1,000–2,000	M: 60–120
(pertechnetate)		stomach	500–1,500	F: 90–180
^{123}I-Iodide	100–400 μCi	thyroid	1,100–4,400	M: 1– 4
		stomach	22– 88	F: 2– 8
^{131}I-Iodide	30 μCi	thyroid	33,000	M: 2.6
				F: 4.2

"It is not possible to compare the calculated absorbed dose in nuclear medicine with diagnostic X-ray exposures for which the exposure conditions are significantly different from those in nuclear medical procedures."

Further differences are:

- Other clinical questions for the procedures.
- The radiation doses for healthy persons and patients are not equal.
- After completing X-ray examination the radiation burden ends for the patient. In nuclear medicine the total radiation dose depends on the remaining radioactivity in the body after the end of examination; the following parameters have an influence on the dose:
  The quantities of the activity of the radiopharmaceutical–the decay scheme of the radioisotope–uptake and distribution in the different organs–long-time retention and in vivo stability of the radiopharmaceutical–organ shape and mass–the age of the patient and, last but not least, the severity of the disease.

e) In conclusion some comments about future prospects of internal dosimetry.
The most significant subjects and goals are accuracy and reliability.
The results of the International Symposium on Radiopharmaceutical Dosimetry held at Oak Ridge in 1980 are /12/:

1. Extrapolation of animal distributions and retention data to man
   When planning animal distribution and retention studies the following experiences should be kept in mind:
   - Mice are not appropriate experimental animals for radiopharmaceuticals, which accumulate in kidneys, the spleen, bone or bone marrow.
   - Rats are neither appropriate for hepatobiliary agents because of the lacking biliary bladder, nor for bone-seekers.
   - Rabbits are not suitable for colloids, labelled red or white blood cells, bone marrow and spleen, imaging agents.

2. Measuring technique for biokinetic data in man
   - A new and modern method is the application of emission computer tomography for quantifying the distribution and retention of radiopharmaceuticals in man.
   - The whole-body counter is always the method of choice conjugated with the transmission computer tomography to determine the effective attenuation.

3. Mathematical, physical and biological models
   - The necessity of using age-dependent phantoms and the suitability of the Reference Man are pointed out.
   - The dosimetric models for the heart and the head must be improved.
   - In estimating the absorbed dose to the fetus from radioisotopes, the lack of specific biokinetic data is even more striking than for adults.

Generally, important progress may be expected over the next years. This will be useful for a better assessment of studies with radioisotopes from the viewpoint of radiation protection and risk estimation and for a better understanding of the distribution and retention of a radiopharmaceutical in the human body.

References

1. *Loevinger, R., M. Bernan:* A schema for absorbed-dose calculations from biologically distributed radionuclides. J. Nucl. Med. **9** (1968), Suppl. No. 1, 7–14; MIRD Pamphlet No. 1, revised 1976.
2. *Brownell, G. L., W. H. Ellett, A. R. Roddy:* Absorbed fractions for photon dosimetry. J. Nucl. Med. **9** (1968), Suppl. No. 3; MIRD-Pamphlet No. 3.
3. *Snyder, W. S., M. R. Ford, G. G. Warner, H. L. Fisher* jr.: Estimates of absorbed fractions for monoenergetic photon sources uniformly distributed in various organs of a heterogeneous phantom. J. Nucl. Med. **10** (1969), Suppl. No. 3; MIRD-Pamphlet No. 5 revised 1978.
4. *Focht, E. F., E. H. Quimby, M. Gershowitz:* Revised average geometric factors for cylinders in isotope dosage. Radiology **85** (1965), 151.
5. *Snyder, W. S., M. R. Ford, G. G. Warner, S. B. Watson:* "S", absorbed dose per unit cumulated activity for selected radionuclides and organs. MIRD-Pamphlet No. 11.
6. *Tautz, M., S. Ertl:* Inkorporationsdosimetrie für die Nuklearmedizin aus der Sicht der MIRD-Empfehlungen. Radiobiol. Radiother. **16** (1976), 65.
7. *Roedler, H. D., A. Kaul, W. Berner, P. Koeppe, D. Glaubitt:* Development of an extended formalism for internal dose calculation and for practical application to several biologically distributed radioelements. In: Assessment of radioactive contamination in man. IAEA, Vienna 1972, IAEA–SM–150/28, p. 515.
8. *Tautz, M., S. Ertl, H. Deckart:* Berechnung der Strahlenbelastung durch Tc-99m-Thioglucose nach der erweiterten MIRD-Konzeption. Radiobiol. Radiother. **20** (1979), 785.
9. *Tautz, M., H. Deckart, S. Ertl:* Ganzkörperretentionsstudien zur Inkorporationsdosimetrie bei Niereninsuffizienz. In: Radioaktive Isotope in Klinik und Forschung, 13th Vol., 2nd part, 1978, p. 575.
10. Task Force on Short-Lived Radionuclides for Medical Applications. J. Nucl. Med. **19** (1978), 107.
11. *Smith, E. M.:* Internal dose calculations for Tc-99m. J. Nucl. Med. **6** (1965), 231.
12. *Roedler, H. D.:* Internal Dosimetry—Present and Future Aspects. International Symposium on Radiopharmaceutical Dosimetry, Oak Ridge 1980.

Index

GPSR Compliance

The European Union's (EU) General Product Safety Regulation (GPSR) is a set of rules that requires consumer products to be safe and our obligations to ensure this.

If you have any concerns about our products, you can contact us on ProductSafety@springernature.com

In case Publisher is established outside the EU, the EU authorized representative is:

Springer Nature Customer Service Center GmbH
Europaplatz 3
69115 Heidelberg, Germany

Batch number: 10761588

Printed by Printforce, the Netherlands